T0186988

# WJEC
# Chemistry
## AS Level

### 2nd Edition

Elfed Charles
Peter Blake
Lindsay Bromley
Jonathan Thomas
Kathryn Foster

Illuminate
Publishing

Published in 2020 by Illuminate Publishing Limited, an imprint of Hodder Education, an Hachette UK Company, Carmelite House, 50 Victoria Embankment, London EC4Y 0DZ

Orders: please contact Hachette UK Distribution, Hely Hutchinson Centre, Milton Road, Didcot, Oxfordshire, OX11 7HH. Telephone: +44 (0)1235 827827. Email: education@hachette.co.uk. Lines are open from 9 a.m. to 5 p.m., Monday to Friday.  You can also order through our website: www.hoddereducation.co.uk

British Library Cataloguing in Publication Data
A catalogue record for this book is available from the British Library

ISBN 978-1-912820-56-6

Printed in the UK

Impression 4

Year 2023

Hachette UK's policy is to use papers that are natural, renewable and recyclable products and made from wood grown in well-managed forests and other controlled sources. The logging and manufacturing processes are expected to conform to the environment regulations of the country of origin.

Every effort has been made to contact copyright holders of material produced in this book. If notified, the publisher will be pleased to rectify any errors or omissions at the earliest opportunity.

This material has been endorsed by WJEC and offers high quality support for the delivery of WJEC qualifications. While this material has been through a WJEC quality assurance process, all responsibility for the content remains with the publisher.

Examination questions are reproduced by permission from WJEC. All exam questions in this book are © WJEC.

Editor: Geoff Tuttle
Design: Nigel Harriss
Layout: Neil Sutton, Cambridge Design Consultants
Original artwork: Greengate Publishing, Tonbridge

Cover image: © Shutterstock

## Acknowledgements

The publisher would like to thank Judith Bonello for her help and advice at key stages in the development process.

### Image credits:

p1 © Shutterstock; p10 Viktoriya/Shutterstock; p16 zhu difeng/Fotofolia; p32 thongsee/Fotofolia; p49 Triff; p50 (l) Triff; p50 (r) MarcelClemens; p59 Shilova Ekaterina; p60 (both t) magnetix; p60 (b) magnetix; p61 (t) Boris 15; p61 (b) magnetix; p62 3Divan; p64 isak55; p67 (l) Lithium, Sodium, Potassium react with water/SquiggleMom/YouTube; p67 (r) Caesium in water/MrGrodskiChemistry/YouTube; p68 (l) ANDREW LAMBERT PHOTOGRAPHY/SCIENCE PHOTO LIBRARY; p68 (r) barium in water/Mostafa El-Sayed/YouTube; p71 Cychr/Creative Commons Attribution 3.0; p73 (all) Irfon Bennett; p76 constantincornel; p82 nikkytok; p83 Photographee.ey; p87 (all) Irfon Bennett; p96 Chepko Danil Vitalevich; p106 Irfon Bennett; p110 Hurst photo; p124 foxbat; p125 (t) Sakarin Sawasdinaka; p125 (m) RGtimeline; p125 (b) majackeka; p126 (t) Diyana Dimitrova; p126 (m) BerndBrueggemann; p126 (b) yuyangc; p127 (t) v.schlichting; p127 (b) Eky Studio; p129 raeva/Fotolia; p130 otterdesign; p142 KDImages/Fotolia; p143 (l) Jeff Zehnder; p143 (r) metriognome; p144 Anticiclo/Fotolia; p153 (l) Dusan Zidar/Fotolia; p153 (r) Pixelspieler; p153 (b) Coprid/Fotolia; p154 markobe/Fotolia; p155 (t) magann/Fotolia; p155 (b) KDImages/Fotolia; p157 adfoto/Fotolia; p162 (l) imagesflyphoto/Fotolia; p162 (r) Photographee.eu/Fotolia; p163 dina777/Photolia; p165 XtravaganT/Photollia; p166 Steve Lovegrove/Photolia; p167 All Canada Photos/Alamy Stock Photo; p174 Africa Stuido/Photolia; p176 WavebreakmediaMicro; p191 R. MACKAY PHOTOGRAPHY, LLC

# Contents

# What this book contains

The contents of this book match the specification for WJEC AS Level Chemistry. It provides you with information and practice examination questions that will help you to prepare for the examinations at the end of the year.

This book addresses:

- The mathematics of chemistry, which will represent a minimum of 20% of your assessment.
- Practical work. The assessment of your practical skills and understanding of experimental chemistry represents a minimum of 15%.
- The three Assessment Objectives required for the WJEC Chemistry course.
- The nature of the examinations.

The book content is clearly divided into the units of this course. These are Unit 1 – The Language of Chemistry, Structure of Matter and Simple Reactions, and Unit 2 – Energy, Rate and Chemistry of Carbon Compounds.

Each chapter covers one topic. The main text of each topic covers all the points from the specification that you need to learn. At the beginning of each chapter is a list of learning objectives highlighting what you need to know and understand and a maths checklist to help you know what maths skills will be required. At the end of each chapter is a set of Test Yourself questions, designed to help you practise for the examinations and to reinforce what you have learned. At the end of each unit, you will find questions selected from WJEC Eduqas examination papers set over the past few years. Answers to all the questions are given at the end of the book. There is also a Periodic table on page 225.

# Marginal features

The margins of each page hold a variety of features to support your learning:

 **Key terms**

**Atomic orbital** is a region in an atom that can hold up to two electrons with opposite spins.

▲ **Key terms** are terms that you need to know how to define. They are highlighted in blue in the body of the text and appear in the Glossary at the back of this book. You will also find other terms in the text in bold type, which are explained in the text, but have not been defined in the margin. The use of key terms is an important feature since examination papers may contain a number of terms that need to be defined.

 **Knowledge check**

(a) Use electrons in boxes to write the electronic configuration of:
 (i) an atom of phosphorus, P.
 (ii) a magnesium ion, $Mg^{2+}$.
(b) Write the electronic configuration in terms of subshells for a chromium atom.

▲ Knowledge check questions are short questions to check your understanding of the subject, allowing you to apply the knowledge that you have acquired. These questions include filling in blanks in a passage, matching terms with phrases specific to the topic under study, and brief calculations. Answers are provided at the back of the book.

 **Study point**

Light is electromagnetic radiation in the range of wavelength corresponding to the visible region of the electromagnetic spectrum.

▲ As you progress through your studies, study points are provided to help you understand and use the knowledge content. In this feature, factual information may be emphasised, or restated to enhance your understanding.

## Stretch & challenge

The mechanism for the hydration of ethene involves electrophilic attack on the π bond by the δ+ H on the water. Try and draw this mechanism using curly arrows.

▲ Stretch and challenge boxes may provide extra information not in the main text, but relevant. They may provide more examples, or questions, but do not contain information that will be tested in an examination.

## Exam tip

Detailed knowledge of the mass spectrometer is not required, however an understanding of how to interpret a mass spectrum is vital.

▲ Exam tips provide general or specific advice to help you prepare for an examination. Read these very carefully.

## Link

Mass spectrometry page 161

▲ Links to other sections of the course are highlighted in the margin, near the relevant text. They are accompanied by a reference to any areas where sections relate to one another. It may be useful for you to use these Links to recap a topic, before beginning to study the current topic.

## Practical check

Preparing a soluble salt by titration is a specified practical task.

When the titration is repeated without an indicator, make sure that when nearing the exact amount needed, the acid is added a drop at a time in order to avoid overshooting.

If you evaporate all of the water then a powder will form instead of crystals.

▲ Occasionally a topic covers an experiment or a practical that is a specified practical task. This feature appears alongside in the margin to highlight its importance and to give you some extra information and hints on understanding it fully.

## Maths tip

When you divide by the smallest to give a whole number ratio, do not cheat! If the ratio is, for example, 1 : 1.5 this is 2 : 3. If the ratio is 1 : 1.67 this is 3 : 5. If your ratio does not simplify to give whole numbers, go back and check – you have made a mistake! One common mistake is to round numbers too soon. Don't do this until the end.

▲ Maths tips provide further explanation about the mathematics described in the text.

# Mathematical skills

Maths crops up quite a lot in AS Level Chemistry (a minimum of 20% of the mark in the exams will depend on your maths skills) so it's really important that you've mastered all the maths skills you'll need before sitting your exams. Don't panic, there is nothing difficult here, you're preparing for a chemistry exam not a maths one! Mathematical requirements are given in Appendix B at the end of the specification course content. The level of understanding is equivalent to Level 2 or GCSE.

You'll be expected to apply a range of maths skills in your exams. These skills are highlighted at the beginning of each chapter and relevant examples to your AS Level Chemistry course are included throughout the book. In addition, further explanations are provided in Maths tips boxes and there is a maths skills chapter on page 191 which gives more explanation and examples of the key mathematical concepts that you need to understand. These features should help you to successfully apply the maths skills that you need.

# Practical work and use of your lab book

This is an important and intrinsic part of the specification and is covered in two ways: firstly as a part of the written papers in which its weighting is at least 15% and, secondly, through direct practical work in the laboratory that will prepare you to deal with the written work. Throughout the AS Level Chemistry course, you will do a lot of practical work and use it to develop many skills. You will record your practical work in a lab book, which is an essential record of what you do. Your lab book is a working document. It is designed to be an ongoing record of the practical tasks you undertake. It is a record of your progress in developing practical techniques, in recording, in making drawings and measurements, plotting graphs, mathematical analysis, evaluating and drawing conclusions.

On pp1–2 of the WJEC lab book is a spreadsheet with a list of the specified practical tasks. The first twelve practicals relate to the AS course. Read the Guidance Notes (pp3–4) in the WJEC lab book every time you do an experiment. It will not take long for you to become familiar with them. They tell you about experimental design, risk assessments, how to display readings and how to plot graphs. They also list aspects of the analysis of your results.

It is important to undertake all twelve practicals, as you may be asked about these, or very similar, experiments in an examination. You should write the date that you did the experiment. There is space for notes and comments. Use this space to remind you of particular issues that occurred to you, so that when you revise for your examinations, you will remember aspects of the experiments that you found significant or challenging at the time.

This book will also help you develop your practical skills and understanding of experimental chemistry. The specified tasks are discussed in detail in the relevant chapters. In addition, advice is given on how to improve a method and how to analyse and evaluate the results.

# Assessment

## Assessment objectives
Examinations test your subject knowledge and the skills associated with how you use that knowledge. These skills are described in Assessment Objectives. Examination questions are written to reflect these objectives, with marks in the proportions shown:

|  | AO1 | AO2 | AO3 |
|---|---|---|---|
| AS Level | 36.2% | 43.8% | 20% |

You must meet these Assessment Objectives in the context of the subject content, which is given in detail in the specification. Your ability to select and communicate information and ideas, clearly using appropriate scientific terminology will be tested within each Assessment Objective. The Assessment Objectives are explained below, with examples of how they are tested. The mark schemes for these questions are on page 214.

## Assessment Objective 1 (AO1)
**Demonstrate knowledge and understanding of scientific ideas, processes, techniques and procedures.**

This AO tests what you know, understand and remember. It is a test of how well you can recall and explain what is relevant. That is why it is essential that you know the content of the specification.

The questions that test this AO are often short-answer questions, using words such as 'state', 'give', 'explain' or 'describe'.

Here are two examples:

### AO1 Demonstrate knowledge
Give the meaning of the word *electronegativity*. [1]

*This question tests AO1 because it asks you to recall factual information.*

### AO1 Demonstrate understanding of scientific ideas
Explain why a change in concentration affects the rate of a reaction. [2]

*This question tests AO1 because it asks for an explanation based on your factual knowledge.*

## Assessment Objective 2 (AO2)
**Apply knowledge and understanding of scientific ideas, processes, techniques and procedures:**
- **in a theoretical context**
- **in a practical context**
- **when handling qualitative data**
- **when handling quantitative data.**

AO2 tests how you use your knowledge and apply it to different situations, in the four possible ways shown above. A question may present you with a situation that you may not have met before, but it will give you enough information so that you can use what you already know to provide an answer.

Make sure you understand the methods of all the experiments you have done; the tests for cations and anions in inorganic chemistry; the tests for functional groups in organic chemistry. Be sure you understand how to do all the calculations needed to process the results.

In testing AO2, a question may have command words such as 'use your knowledge of…', 'calculate' or 'explain…'.

Here are four examples:

### AO2 In a theoretical context
Use ideas that you have studied in your Chemistry course to comment on and explain the following observation.

When dilute sulfuric acid is added to aqueous magnesium chloride no visible change occurs but when it is added to aqueous barium chloride a white precipitate is observed.

Include an equation for any reaction that you observe. [3]

*'Use ideas that you have studied' indicates that you should use your theoretical knowledge and understanding to explain a chemical observation.*

### AO2 In a practical context
A compound is known to be one of butan-2-ol, $CH_3CH_2CH(OH)CH_3$, 2-methylpropanoic acid, $CH_3CH(CH_3)COOH$, and 3-hydroxybutanoic acid, $CH_3CH(OH)CH_2COOH$.

Choose **two** chemical tests that will enable you to determine which one it is.

Complete the table below.

Give the reagent(s) for each test and the observation expected for a positive result. For each test put a tick (✓) in the box to show the compound that gives a positive result and a cross (✗) for those that do not. [5]

| Reagent(s) | Observation expected for positive result | butan-2-ol | 2-methyl-propanoic acid | 3-hydroxy-butanoic acid |
|---|---|---|---|---|
|  |  |  |  |  |
|  |  |  |  |  |

*This is AO2 because you are asked to show how you would use chemical tests in a practical context to identify a compound.*

## AO2 When handling qualitative data

Two reactions of organic compound A are shown below.

$$B \xleftarrow{\text{conc } H_2SO_4} A \xrightarrow{Cr_2O_7^{2-}/H^+} C$$

Compound **B** is a straight-chain hydrocarbon with the formula $C_4H_8$.

Draw the displayed formulae of two possible isomers of **A**. [2]

*Analytical data has been provided which gives information about the functional group of an organic compound. It is non numerical and so is qualitative.*

## AO2 When handling quantitative data

Calculate the volume, in $cm^3$, that $2.54 \times 10^{-3}$ mol of nitrogen occupy at a temperature of 120 °C and a pressure of 101 kPa. [4]

*The information given in this question is quantitative and you are asked to apply mathematical procedures to calculate a volume.*

## Assessment Objective 3 (AO3)

**Analyse, interpret and evaluate scientific information, ideas and evidence to:**

- **make judgements and reach conclusions**
- **develop and refine practical design and procedures.**

The AO3 marks on a paper are awarded for developing and refining practical design and procedures, for making judgements and for drawing conclusions. You may be asked to criticise a method or analysis or be asked how to improve aspects of it. You may be asked to design a method to test a particular hypothesis or why you might apply a particular test. You may be asked to interpret a test or draw a conclusion from evidence presented to you.

Examination questions will give information in novel situations. As with AO2 questions, you may be presented with unfamiliar scenarios, but you will be tested on how well you use your knowledge to understand and interpret them.

For all the experiments you have done, make sure you know how to improve the accuracy of your method and the repeatability of readings.

AO3 questions tend to use words such as 'evaluate', 'suggest', 'justify', 'improve' or 'design' and often ask you to manipulate experimental data.

Here are two examples:

## AO3 Make judgements and reach conclusions

Chlorine forms molecules and ions with other halogens. Two examples are $[ClF_2]^+$ and $[ClF_2]^-$

A student said that since these ions have a different charge their shapes must be different. Is he correct? Justify your answer by using VSEPR theory. [3]

*This question is AO3 because it asks you to draw a conclusion about the shapes of these unfamiliar ions and to make a judgement on the student's statement.*

## AO3 develop and refine practical design and procedures

Kathryn is asked to determine the enthalpy change of solution of ammonium nitrate, $NH_4NO_3$, using the following method:

- Weigh 7.40 g of ammonium nitrate
- Weigh 50.0 g of water in a polystyrene cup
- Record the temperature of the water
- Add the ammonium nitrate and stir until it has all dissolved
- Record the minimum temperature of the solution

She noted that the temperature fell from 20.0 °C to 10.8 °C.

Kathryn's results gave a lower value than the literature value.

Suggest two improvements to her method and explain how they would lead to a more accurate value. [4]

*This question is AO3 because asking for a method to improve the accuracy of the results is asking how to refine the experimental procedure.*

# Examinations

Two units will be examined at AS Level. The assessment is summarised in the table below:

| Paper | Unit 1 The Language of Chemistry, Structure of Matter and Simple Reactions | Unit 2 Energy, Rate and Chemistry of Carbon Compounds |
|---|---|---|
| Topics covered | Topic 1.1–1.7 | Topic 2.1–2.8 |
| % of the AS qualification | 50% | 50% |
| Length of exam | 1 hour 30 minutes | 1 hour 30 minutes |
| Marks available | 80 | 80 |
| Question types | Short answer<br>Structured<br>Calculation<br>Extended prose | Short answer<br>Structured<br>Calculation<br>Extended prose |
| Practical | A minimum of 15% of the marks across both papers will be awarded for practical knowledge and understanding | |
| Maths | A minimum of 20% of the marks across both papers will be awarded for mathematics at Level 2 | |

## Examination questions

As well as being able to recall facts, write equations, name structures and describe their functions, you need to appreciate the underlying principles of the subject and understand associated concepts and ideas. In other words, you need to develop skills so that you can apply what you have learned, even to situations not previously encountered.

You will be expected to answer different styles of question, for example:

- Short answer questions – these require a brief answer, such as a formula, an equation, an observation or a definition.

- Structured questions – these are in several parts, often about a common theme. They can become more difficult as you work your way through. Structured questions can be short, requiring a brief response, or may include the opportunity for extended writing. The number of lined spaces and the mark allocation at the end of each part question are there to help you. They indicate the length of answer expected. If three marks are allocated then you must give at least three separate points.

- Calculations – you may be asked to convert between units, rearrange formulae and equations, analyse and evaluate numerical data or inter-convert numerical data and graph form. Although these mathematical questions account for a minimum of 20% of the total mark, this does not mean 20% in each paper. Unit 1 will carry a greater percentage of maths marks than Unit 2.

- Extended prose question – each paper will contain one six-mark question which requires extended prose for its answer. Often, candidates rush into such questions. You should take time to read it carefully to discover exactly what the examiner requires in the answer, and then construct a plan. This will

not only help you organise your thoughts logically but will also give you a checklist to which you can refer when writing your answer. In this way you will be less likely to repeat yourself, wander off the subject or omit important points. Examiners do not award marks for individual items of information, but a more holistic approach using a banded level of response is used.

5–6 marks will be awarded if you include most of the relevant factual information with clear scientific reasoning. The piece of writing should answer the question directly, using well-constructed sentences and suitable chemical terminology.

3–4 marks will be awarded if there are some factual omissions.

1–2 marks will be awarded if only a few valid points are made, with little use of scientific vocabulary.

0 marks will be awarded if the question is not attempted or no relevant points are made.

Since practical work is an essential part of chemistry, questions assessing your practical knowledge and understanding can be part of any style of question, e.g. giving an observation in a short answer question, explaining how refining an experiment can improve the result as part of a structured question or planning an experiment in an extended prose question.

## Command words

Command words are just the part of the question that tell you what to do. You'll find answering the exam questions much easier if you understand exactly what they mean. The most common command words are summarised in the table below:

| Command words | What to do |
|---|---|
| Give/Name/State | Give a one-word answer or a short sentence with no explanation. |
| State what you see/Describe an observation | Write about what you would expect to happen in a reaction, e.g. effervescence seen, a white precipitate forms. |
| Describe or outline an experiment | Give a step-by-step account of what is taking place. Include any equipment you would use, the reagents required and any reaction conditions. |
| Explain | Give chemical reasons for something. |
| Suggest/Predict | There may not be a definite answer to the question. Use your chemical knowledge to put forward a sensible idea to what the answer might be. |
| Compare | Give the similarities and/or differences between two things. Make an explicit comparison in each sentence. |
| Deduce | Use the information provided to answer the question. |
| Calculate | Use the information provided and your mathematical knowledge to work out the answer. |
| Justify | You will be given a statement for which you should use your chemical knowledge as evidence in support or to the contrary. Draw a conclusion as to whether the initial statement can be accepted. |
| Write an equation | You will need to know the formulae of the reactants and products. Also, you must ensure that the equation is balanced. |

Not all of the questions will have command words. Some questions will also ask you to answer 'using the information given', e.g. a graph, table, equation, etc. If that's the case, you must use the information given and you may need to refer to it in your answer to get all the marks.

## Exam Data Booklet

When you sit your exams as well as the exam paper, you'll be given a data booklet. In it you'll find some very useful information to help you with your exam, including:

- Avogadro's constant and the gas constant ($R$)
- Molar gas volumes at 273 K / 298 K and 1 atm
- Planck's constant ($h$) and the speed of light ($c$)
- The specific heat capacity of water ($c$)
- The relationship between $m^3$ and $cm^3$, K and °C, atm and Pa.
- Infrared absorption values of some functional groups
- $^{13}C$ NMR chemical shifts of some types of carbon
- $^1H$ NMR chemical shifts of some types of proton
- A copy of the periodic table

Make sure you use the information from the periodic table in the data booklet, even if you think it's slightly different from something you've seen elsewhere. The information in the data booklet is what the examiners use to mark the exam papers.

# How to maximise your score

We all vary in speed and natural ability but by attacking the challenge of AS Level Chemistry in the right way the best possible outcome can be achieved. There are two parts to achieving your maximum score – exam revision and exam strategy.

## Exam revision

This is the hard work requiring private concentration. It is essential that you learn facts, concepts and information by heart. Understanding more complex concepts requires you to have a body of knowledge. The key is to get a good, clear set of notes and work on them small amounts at a time until you know them. Make lists, draw and annotate diagrams from memory; read your notes and repeat them out loud; ask your friends and relatives to test you.

Give yourself plenty of time so that you can return to each topic after an interval of time to ensure that you still have it mastered. This may take more than one return trip. It is known that the unconscious mind continues to work and sort learned material. Last minute cramming is not as efficient.

There is no substitute for work. Top sports players practise thousands of serves or goal kicks. Geraint Thomas did many very long training sessions on his bicycle – winners work hard!

## Exam strategy

- Make sure you have all the correct equipment needed for your exam.

- Plan your time; how long you spend on each question is really important. Try to leave some time at the end of the exam to check your answers. Some questions will require lots of work for only a few marks, but other questions can be answered much more quickly. Don't spend ages struggling with questions that are only worth a couple of marks, move on and come back to them at the end of the exam.

- Read the question carefully to work out what you are being asked to do. Are there any command words in the question? What is the mark allocation? How many lines have you been given to write on? All of these will give you an indication of what is required in your answer.

If it's a calculation, always show your working. Marks will be allocated for individual steps, not just the final answer.

If it's an extended prose question (QER), plan your answer so that you structure your response logically showing how the points you make follow on from each other.

- Diagrams and graphs can earn marks more easily than written explanations, but only if they are carefully drawn. Diagrams do not need to be works of art, but make sure that they are clear and fully labelled. When drawing graphs, make sure that you use a sensible scale and label the axes with quantities and units.

- Check your answers. If a question is worth three marks, have you made three distinct points? If it's a calculation, pay particular attention to using the correct units. Does the answer make sense? Check each stage of your working.

# Formulae and equations

In chemistry, each element has a symbol. A symbol is a letter or two letters which stand for one atom of the element. Formulae are written for compounds. Formulae consist of the symbols of the elements present and numbers that show the ratio in which the atoms are present. Using symbols and formulae enables you to write equations for chemical reactions.

Atoms are neither created nor destroyed in a chemical reaction, therefore when you write a chemical equation, the same number of atoms of each element must be present on each side of the equation. This is done by balancing the equation.

## Topic contents

You should be able to demonstrate and apply knowledge and understanding of:

- Formulae of common compounds and common ions and how to write formulae for ionic compounds.
- Oxidation numbers of atoms in a compound or ion.
- How to construct balanced chemical equations, including ionic equations, with appropriate state symbols.

### Maths skills ≫

- Use ratios to write formulae and construct and balance equations.

You should have covered formulae and equations at GCSE level, but this section gives a recap on the minimum knowledge required at AS level. It also includes a section on oxidation numbers which will be new to you.

# Formulae of compounds and ions

The formula of a compound is a set of symbols and numbers. The symbols say what elements are present and the numbers give the ratio of the numbers of atoms of the different elements in the compound.

The compound carbon dioxide has the formula $CO_2$. It contains two oxygen atoms for every carbon atom. The formula for ethanol is $C_2H_5OH$. It contains two carbon atoms and six hydrogen atoms for every oxygen atom. These compounds consist of molecules in which the atoms are bonded covalently. To show two molecules you write $2C_2H_5OH$. The 2 in front of a formula multiplies everything after it. Therefore, in $2C_2H_5OH$, there are 4 C, 12 H and 2 O atoms, a total of 18 atoms.

For advanced level chemistry you will need to know the formulae of a wide range of compounds. The table below gives a list of the formulae of some common compounds. Since many compounds do not consist of molecules but consist of ions and form through ionic bonding, the list contains both ionic and covalent compounds.

**‹ Link ›**

Covalent and ionic bonding pages 50–51

| Name | Formula | Name | Formula |
|---|---|---|---|
| Water | $H_2O$ | Sodium hydroxide | NaOH |
| Carbon dioxide | $CO_2$ | Sodium chloride | NaCl |
| Sulfur dioxide | $SO_2$ | Sodium carbonate | $Na_2CO_3$ |
| Methane | $CH_4$ | Sodium hydrogencarbonate | $NaHCO_3$ |
| Hydrochloric acid | HCl | Sodium sulfate | $Na_2SO_4$ |
| Sulfuric acid | $H_2SO_4$ | Copper(II) oxide | CuO |
| Nitric acid | $HNO_3$ | Copper(II) sulfate | $CuSO_4$ |
| Ethanoic acid | $CH_3COOH$ | Calcium hydroxide | $Ca(OH)_2$ |
| Ammonia | $NH_3$ | Calcium carbonate | $CaCO_3$ |
| Ammonium chloride | $NH_4Cl$ | Calcium chloride | $CaCl_2$ |

The compound calcium chloride is composed of calcium ions, $Ca^{2+}$, and chloride ions, $Cl^-$. There are twice as many chloride ions as calcium ions, so the formula is $CaCl_2$. This is not a molecule of calcium chloride but a formula unit of calcium chloride. For an ionic compound the total number of positive charges must equal the total number of negative charges in one formula unit of the compound.

The table below gives the formulae for common ions that you need to learn.

| Positive ions | | Negative ions | |
|---|---|---|---|
| Name | Formula | Name | Formula |
| Ammonium | $NH_4^+$ | Bromide | $Br^-$ |
| Hydrogen | $H^+$ | Chloride | $Cl^-$ |
| Lithium | $Li^+$ | Fluoride | $F^-$ |
| Potassium | $K^+$ | Iodide | $I^-$ |
| Sodium | $Na^+$ | Hydrogencarbonate | $HCO_3^-$ |
| Silver | $Ag^+$ | Hydroxide | $OH^-$ |
| Barium | $Ba^{2+}$ | Nitrate | $NO_3^-$ |
| Calcium | $Ca^{2+}$ | Oxide | $O^{2-}$ |
| Magnesium | $Mg^{2+}$ | Sulfide | $S^{2-}$ |
| Copper(II) | $Cu^{2+}$ | Carbonate | $CO_3^{2-}$ |
| Iron(II) | $Fe^{2+}$ | Sulfate | $SO_4^{2-}$ |
| Iron(III) | $Fe^{3+}$ | Phosphate | $PO_4^{3-}$ |
| Aluminium | $Al^{3+}$ | | |

Note that non-metals change to end in -ide, but if non-metals combine with oxygen to form negative ions, the negative ion starts with the non-metal and ends in -ate (apart from hydroxide).

**Knowledge check 1**

How many atoms of each element are present in:

(a) $P_4O_{10}$

(b) $2Al(OH)_3$?

**Knowledge check 2**

How many oxygen atoms are present in $3Fe(NO_3)_3$?

**Knowledge check 3**

Name the following compounds:

(a) $Na_2SO_4$

(b) $Ca(HCO_3)_2$

(c) $CuCl_2$

**4 Knowledge check**

Give the formula for:

(a) Aluminium oxide

(b) Potassium carbonate

(c) Ammonium sulfate.

The formula for ionic compounds can be calculated by following these steps:

1. Write the symbols of the ions in the compound.
2. Balance the ions so that the total of the positive ions and negative ions adds to zero. (The compound itself must be neutral.)
3. Write the formula without the charges and put the number of ions of each element as a small number following and below the element symbol.

**Worked examples**

**Example 1**

Magnesium oxide

1. The ions are $Mg^{2+}$ and $O^{2-}$.
2. To make the total charge zero, we need one $Mg^{2+}$ ion for every $O^{2-}$ ion (+2 −2 = 0).
3. Formula is MgO (the '1' does not need to be included).

**Example 2**

Sodium sulfide

1. The ions are $Na^+$ and $S^{2-}$.
2. To make the total charge zero, we need two $Na^+$ ions for every $S^{2-}$ (+1 +1 −2 = 0) i.e. $Na^+ Na^+ S^{2-}$.
3. Formula is $Na_2S$.

**Example 3**

Calcium nitrate

1. The ions are $Ca^{2+}$ and $NO_3^-$.
2. Two $NO_3^-$ ions are needed to balance the charge on one $Ca^{2+}$ ion (−1 −1 +2 = 0) i.e. $Ca^{2+} NO_3^- NO_3^-$.
3. Formula is $Ca(NO_3)_2$ (note the use of a bracket around the $NO_3$ before adding the 2).

**Key term**

**Oxidation number** is the number of electrons that need to be added to (or taken away from) an element to make it neutral.

**Study point**

Sometimes the term oxidation state is used instead of oxidation number. The only difference is that we say for $Fe^{2+}$, for example, the oxidation number of Fe is +2 but the oxidation state of this ion is written as Fe(II).

**Study point**

It is useful to use simple steps in determining oxidation numbers.

E.g. What is the oxidation number of nitrogen in the $NO_3^-$ ion?

Step 1 The oxidation number of each oxygen is −2.

Step 2 The total for $O_3$ is −6.

Step 3 The overall charge on the ion is −1.

Step 4 The oxidation number of nitrogen is (−1) − (−6) = +5.

# Oxidation numbers

As you have seen on the previous pages, there are differences in the ratios of the atoms that combine together to form compounds, e.g. $H_2O$ and HCl. A method of expressing the combining power of elements is **oxidation number**. The oxidation number of an element is the number of electrons that need to be added to (or taken away from) an element to make it neutral.

For example, the iron(II) ion, $Fe^{2+}$, needs the addition of two electrons to make a neutral atom, therefore it has the oxidation number +2. The chloride ion, $Cl^-$, needs to lose an electron to make a neutral atom; therefore it has the oxidation number −1.

The use of oxidation numbers can be extended to covalent compounds. Some elements are assigned positive oxidation numbers and others are assigned negative oxidation numbers in accordance with certain rules, given in the following table.

| Rule | Example |
|------|---------|
| The oxidation number of an uncombined element is zero. | Metallic copper, Cu: oxidation number 0<br><br>Oxygen gas, $O_2$: oxidation number 0. |
| The sum of the oxidation numbers in a compound is zero.<br><br>In an ion the sum equals the overall charge. | In $CO_2$ the sum of the oxidation numbers of carbon and oxygen is 0.<br><br>In $NO_3^-$ the sum of the oxidation numbers of nitrogen and oxygen is $-1$. |
| In compounds the oxidation numbers of Group 1 metals is $+1$ and Group 2 metals is $+2$. | In $MgBr_2$ the oxidation number of magnesium is $+2$ (oxidation number of each bromine is $-1$). |
| The oxidation number of oxygen is $-2$ in compounds except with fluorine or in peroxides (and superoxides). | In $SO_2$ the oxidation number of each oxygen is $-2$ (oxidation number of sulfur is $+4$).<br><br>In $H_2O_2$ the oxidation number of oxygen is $-1$ (oxidation number of hydrogen is $+1$). |
| The oxidation number of hydrogen is $+1$ in compounds except in metal hydrides. | In HCl the oxidation number of hydrogen is $+1$ (oxidation number of chlorine is $-1$).<br><br>In NaH the oxidation number of hydrogen is $-1$ (oxidation number of sodium is $+1$). |
| In chemical species with atoms of more than one element, the most electronegative element is given the negative oxidation number. | In $CCl_4$, chlorine is more electronegative than carbon, so the oxidation number of each chlorine is $-1$ (oxidation number of carbon is $+4$). |

Oxidation numbers are used in redox reactions (reactions where both **red**uction and **ox**idation take place) to show which species is oxidised and which one is reduced. If the oxidation number of a species increases, it is oxidised; if the oxidation number decreases, it is reduced.

Oxidation numbers are used to name compounds unambiguously, e.g. potassium, nitrogen and oxygen can combine to give two different compounds, $KNO_3$ and $KNO_2$. Since the oxidation number of potassium is $+1$ and that of oxygen is $-2$, the oxidation number of nitrogen must be $+5$ in $KNO_3$ and $+3$ in $KNO_2$. Therefore $KNO_3$ is called potassium nitrate(V) and $KNO_2$ is called potassium nitrate(III).

# Chemical and ionic equations

Chemical equations are written to sum up what happens in a chemical reaction.

Since atoms are neither created nor destroyed in a chemical reaction, there must be the same number of atoms of each element on each side of the chemical equation.

The steps in writing a balanced chemical equation are:

1. Write a word equation for the reaction (optional).
2. Write the symbols and formulae for the reactants and products (make sure that all formulae are correct).
3. Balance the equation by multiplying formulae if necessary (never change a formula).
4. Check your answer.
5. Add state symbols (if required).

The state symbols used are: (s) for solid, (l) for liquid, (g) for gas. A solution in water is described as aqueous, so (aq) is used for a solution.

## Knowledge check 5

What is the oxidation number of:
(a) nitrogen in $NH_3$
(b) phosphorus in $P_4$
(c) manganese in $MnO_4^-$
(d) chromium in $K_2Cr_2O_7$?

## Exam tip

Always write an oxidation number with the sign of the charge first, followed by the number, e.g. the oxidation number of sulfur in $SO_4^{2-}$ is $+6$. Writing 6+ is incorrect since an $S^{6+}$ ion does not exist.

## Link

Electronegativity page 52
Oxidation and reduction page 67

## Exam tip

All the elements in the list H O F Br I N Cl are diatomic molecules. You could create your own mnemonic to remember them in the exam.

## Knowledge check 6

Balance the following equations
(a) $SO_2 + O_2 \longrightarrow SO_3$
(b) $Fe_2O_3 + CO \longrightarrow Fe + CO_2$
(c) $Al + HCl \longrightarrow AlCl_3 + H_2$
(d) $HI + H_2SO_4 \longrightarrow I_2 + H_2O + H_2S$

**Worked example**

Sodium carbonate reacts with dilute hydrochloric acid to give carbon dioxide and a solution of sodium chloride.

Write a balanced chemical equation including state symbols for this reaction.

$$\text{sodium carbonate} + \text{hydrochloric acid} \longrightarrow \text{sodium chloride} + \text{carbon dioxide} + \text{water}$$

Writing the formulae gives:

$$Na_2CO_3 + HCl \longrightarrow NaCl + CO_2 + H_2O$$

Number of atoms on L.H.S. = 2Na + 1C + 3O + 1H + 1Cl

Number of atoms on R.H.S. = 1Na + 1C + 3O + 2H + 1Cl

Start by balancing Na atoms, so multiply NaCl on the R.H.S. by 2:

$$Na_2CO_3 + HCl \longrightarrow 2NaCl + CO_2 + H_2O$$

Number of atoms on L.H.S. = 2Na + 1C + 3O + 1H + 1Cl

Number of atoms on R.H.S. = 2Na + 1C + 3O + 2H + 2Cl

Next balance H atoms by multiplying HCl on the L.H.S. by 2:

$$Na_2CO_3 + 2HCl \longrightarrow 2NaCl + CO_2 + H_2O$$

Number of atoms on L.H.S. = 2Na + 1C + 3O + 2H + 2Cl

Number of atoms on R.H.S. = 2Na + 1C + 3O + 2H + 2Cl

Equation is now balanced. Add state symbols.

$$Na_2CO_3(s) + 2HCl(aq) \longrightarrow 2NaCl(aq) + CO_2(g) + H_2O(l)$$

### Stretch & challenge

Balance the following equation:

$Cu + HNO_3 \longrightarrow Cu(NO_3)_2 + NO + H_2O.$

## Ionic equations

Many reactions involve ions in solutions. However, in these reactions not all of the ions take part in any chemical change. An ionic equation may help to show what is happening.

Ionic equations are frequently used for displacement and precipitation reactions.

### 7 Knowledge check

When a solution of sodium sulfate is added to a solution of barium chloride, a white precipitate of barium sulfate forms. Write an ionic equation, including state symbols, for this reaction.

**Worked example 1**

When zinc powder is added to copper(II) sulfate solution, copper is displaced and a red-brown deposit is formed on the zinc.

The chemical equation for the reaction is:

$$Zn(s) + CuSO_4(aq) \longrightarrow ZnSO_4(aq) + Cu(s)$$

Writing out all of the ions gives:

$$Zn(s) + Cu^{2+}(aq) + SO_4^{2-}(aq) \longrightarrow Zn^{2+}(aq) + SO_4^{2-}(aq) + Cu(s)$$

There is repetition here. The $SO_4^{2-}(aq)$ ions have not taken part in any chemical change at all. They have been present unchanged throughout. They are called spectator ions and are left out of the ionic equation, which is written:

$$Zn(s) + Cu^{2+}(aq) \longrightarrow Zn^{2+}(aq) + Cu(s)$$

An ionic equation provides a shorter equation which focuses attention on the changes taking place.

**Worked example 2**

When a solution of sodium hydroxide is added to a solution of magnesium chloride a white precipitate forms.

The chemical equation for the reaction is:

$$2NaOH(aq) + MgCl_2(aq) \longrightarrow 2NaCl(aq) + Mg(OH)_2(s)$$

Writing out all of the ions gives:

$$2Na^+(aq) + 2OH^-(aq) + Mg^{2+}(aq) + 2Cl^-(aq) \longrightarrow 2Na^+(aq) + 2Cl^-(aq) + Mg(OH)_2(s)$$

The $Na^+(aq)$ ions and the $Cl^-(aq)$ ions do not change during the reaction. They are spectator ions and can be omitted, giving the ionic equation:

$$Mg^{2+}(aq) + 2OH^-(aq) \longrightarrow Mg(OH)_2(s)$$

**‹ Link ›**

Solubility of Group 2 cations page 69.

# Test yourself

1. Radium carbonate has the formula $RaCO_3$. Write the formula of radium hydroxide. [1]

2. State the oxidation number of chromium in $CrO_2Cl_2$ [1]

3. Magnetite ore can be reduced by carbon monoxide in a blast furnace to produce iron as part of steel production.

   Balance the equation for the reduction of magnetite.

   $$Fe_3O_4 + CO \longrightarrow Fe + CO_2$$ [1]

4. When ethanol, $C_2H_5OH$, is burnt in air, the only products are carbon dioxide and water.

   Balance the equation for this reaction.

   $$C_2H_5OH + O_2 \longrightarrow CO_2 + H_2O$$ [1]

5. When calcium is added to cold water, calcium hydroxide and hydrogen form.

   Write the balanced chemical equation for this reaction. [1]

6. Write a balanced equation for the reaction between $SiCl_4$ and water to form $SiO_2$ and HCl only. [1]

7. Lead phosphate is precipitated when aqueous ammonium phosphate is mixed with aqueous lead nitrate.

   Write the balanced chemical equation for this reaction. [1]

8. An oxide of nitrogen reacts with water to form nitric acid as one of the products:

   $$NO_2(g) + H_2O(l) \longrightarrow HNO_3(g) + NO(g)$$

   (a) Balance the equation above. [1]

   (b) Give the oxidation numbers of nitrogen in all three nitrogen species. [2]

9. When an aqueous solution of calcium hydroxide is added to an aqueous solution of sodium carbonate a white precipitate of calcium carbonate is seen.

   Write the **ionic** equation for this reaction. Include the relevant state symbols in the equation. [1]

## 1.2

# Basic ideas about atoms

Chemistry is the study of how matter behaves. We know that all matter is made up of very small particles called atoms. The idea of atoms was put forward by the Greeks in the fifth century BCE but it was not until the nineteenth and early twentieth century that scientists showed that matter is made up of atoms and atoms have an internal structure comprising protons, neutrons and electrons.

Protons and neutrons are made from quarks, and electrons belong to the lepton particle family. This unit looks at protons, neutrons and electrons. It shows what happens when an unstable atom splits to form smaller particles and how ionisation energies and emission spectra provide evidence for electronic configuration.

## Topic contents

You should be able to demonstrate and apply knowledge and understanding of the:

- Nature of radioactive decay and the resulting changes in atomic number and mass number (including positron emission and electron capture).
- Behaviour of α-, β- and γ-radiation in electric and magnetic fields and their relative penetrating power.
- Half-life of radioactive decay.
- Adverse consequences for living cells of exposure to radiation and use of radioisotopes in many contexts, including health, medicine, radio-dating, industry and analysis.
- Significance of standard molar ionisation energies of gaseous atoms and their variation from one element to another.
- Link between successive ionisation energy values and electronic structure.
- Shapes of s- and p-orbitals and the order of s-, p- and d-orbital occupation for elements 1–36.
- Origin of emission and absorption spectra in terms of electron transitions between atomic energy levels.
- Atomic emission spectrum of the hydrogen atom.
- Relationship between energy and frequency ($E = hf$) and that between frequency and wavelength ($f = c/\lambda$).
- Order of increasing energy of infrared, visible and ultraviolet light.
- Significance of the frequency of the convergence limit of the Lyman series and its relationship with the ionisation energy of the hydrogen atom.

### Maths skills ≫

- Use fractions and percentages in calculations.
- Make use of appropriate units in calculations.
- Use an appropriate number of significant figures.
- Substitute numerical values into algebraic equations.
- Change the subject of an equation.

Atomic structure is not specifically mentioned in the specification. However, since all learners are expected to demonstrate knowledge and understanding of standard content covered at GCSE level, pages 17–18 give a recap on the minimum knowledge that is required about the structure of the atom and how elements and ions are represented.

# Atomic structure

Atoms are made up of three fundamental particles: the proton, the neutron and the electron.

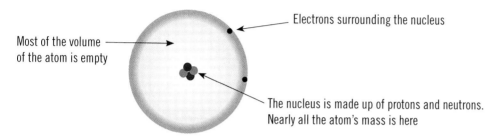

Electrons surrounding the nucleus

Most of the volume of the atom is empty

The nucleus is made up of protons and neutrons. Nearly all the atom's mass is here

The masses and charges of these particles are very small and so are inconvenient, therefore we call the mass of a proton 1, its charge +1 and we describe the other particles relative to these values.

| Particle | Relative mass | Relative charge |
| --- | --- | --- |
| proton | 1 | +1 |
| neutron | 1 | 0 |
| electron | Negligible (1/1840) | −1 |

An atom is electrically neutral because the number of negative electrons surrounding the nucleus equals the number of positive protons in the nucleus.

# Representing elements and ions

All atoms of the same element contain the same number of protons. The number of protons in the nucleus of an atom determines the element to which the atom belongs and is known as the **atomic number**.

It is also useful to have a measure for the total number of particles in the nucleus of an atom. This is called the **mass number**.

The full symbol for an element incorporates the atomic number, mass number and symbol

e.g.

mass number $\longrightarrow$ 23

**Na** $\longleftarrow$ symbol

atomic number $\longrightarrow$ 11

Atoms of the same element are not all identical. They always have the same number of protons, but they can have different numbers of neutrons. Such atoms are called **isotopes**. Most elements exist naturally as two or more different isotopes. For example, chlorine consists of two isotopes, one having a mass number of 35 and one having a mass number of 37 or $^{35}_{17}\text{Cl}$ and $^{37}_{17}\text{Cl}$.

A particle where the number of electrons does not equal the number of protons is no longer an atom but is called an **ion** and has an electrical charge.

**Link**

Ionic bonding page 50

**2** **Knowledge check**

State the number of protons and electrons in

(a) $^{131}I^-$

(b) $^{25}Mg^{2+}$.

**Stretch & challenge**

Nuclei contain protons packed together in a very small space. Why do nuclei not fly apart?

**Exam tip**

If you are given an element's mass number and symbol, use the periodic table to find its atomic number. Remember it might be an isotope so the mass number might be different from that in the periodic table.

**Key terms**

$\alpha$-**particles** have a nucleus of 2 protons and 2 neutrons, therefore positively charged.

$\beta$-**particles** are fast moving electrons, therefore negatively charged.

$\gamma$-**rays** are high energy electromagnetic radiation, therefore no charge.

**Stretch & challenge**

$\beta$ particles can be considered as being formed when a neutron changes into a proton, i.e.

$$^1_0n \longrightarrow {}^1_1p + {}^0_{-1}\beta.$$

If a neutral atom loses one or more electrons it forms a positive ion or cation,

e.g. $$Na \longrightarrow Na^+ + e^-$$

If a neutral atom gains one or more electrons it forms a negative ion or anion,

e.g. $$Cl + e^- \longrightarrow Cl^-$$

In both examples the number of protons has not changed but the number of electrons has.

The number of electrons in $Na^+$ is 10 (atomic number − charge on ion).

The number of electrons in $Cl^-$ is 18 (atomic number + charge on ion).

# Radioactivity

## Types of radioactive emission and their behaviour

Some isotopes are unstable and split up to form smaller atoms. The nucleus divides and sometimes protons, neutrons and electrons fly out. The process is called radioactive decay and the element is said to be radioactive. Radioactive isotopes have unstable nuclei and they give off three types of radiation: **alpha ($\alpha$)**, **beta ($\beta$)** and **gamma ($\gamma$)**.

**Alpha particles** consist of two protons and two neutrons and are therefore helium nuclei. They are the least penetrating of the three types of radiation and are stopped by a thin sheet of paper or even a few centimetres of air.

**Beta particles** consist of streams of high-energy electrons and are more penetrating. They can travel through up to 1 m of air but are stopped by a 5 mm thick sheet of aluminium.

**Gamma rays** are high-energy electromagnetic waves and are the most penetrating of the three radiations. They can pass through several centimetres of lead or more than a metre of concrete.

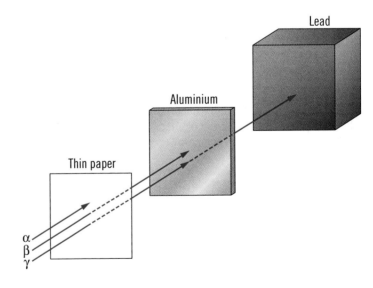

▲ The penetrating powers of radiation

When alpha, beta and gamma radiation pass through matter they tend to knock electrons out of atoms, ionising them. Alpha particles are strongly ionising because they are large, relatively slow moving and carry two positive charges. On the other hand, gamma rays are only weakly ionising.

Ionisation involves a transfer of energy from the radiation passing through the matter to the matter itself. As the alpha particle is the most strongly ionising of the radiations, this transfer happens most rapidly and so they are the least penetrating. Conversely, since gamma rays are the least ionising they are the most penetrating of the radiations.

When alpha, beta and gamma radiations pass through an electric field, gamma rays are undeflected, while alpha particles are deflected towards the negatively charged plate and beta particles towards the positive plate.

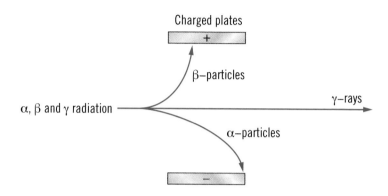

▲ The effect of electric field on radiation

Magnetic fields have a similar effect on alpha, beta and gamma radiation. When a charged particle cuts through a magnetic field it experiences a force referred to as the motor effect.

Alpha particles are deflected by a magnetic field, confirming that they must carry a charge. The direction of deflection (which can be determined by Fleming's left-hand rule) demonstrates that they must be positively charged.

Beta particles are deflected by a magnetic field in an opposite direction to alpha particles confirming they must hold a charge opposite to alpha particles.

Gamma rays are unaffected by a magnetic field. This shows gamma rays are uncharged as they do not experience a force when passing through the lines of a magnetic field.

# Effect on mass number and atomic number

$\alpha$ and $\beta$ particle emissions result in the formation of a new nucleus with a new atomic number therefore the product is a different element.

When an element emits an $\alpha$ particle its mass number decreases by 4 and its atomic number decreases by 2.

$$^{238}_{92}\text{U} \longrightarrow {}^{234}_{90}\text{Th} + {}^{4}_{2}\alpha$$

The product is two places to the left in the periodic table.

When an element emits a $\beta$ particle its mass number is unchanged and its atomic number increases by 1.

$$^{14}_{6}\text{C} \longrightarrow {}^{14}_{7}\text{N} + {}^{0}_{-1}\beta$$

The product is one place to the right in the periodic table.

A process of inverse beta decay can also occur. This is known as **electron capture**. In the process of electron capture, one of the orbital electrons is captured by a proton in the nucleus, forming a neutron (and emitting an electron neutrino, $\nu_e$.)

### ⟩⟩ Study point

In equations:
$^{4}_{2}\text{He}^{2+}$ is acceptable for $^{4}_{2}\alpha$.
$^{0}_{-1}\text{e}$ is acceptable for $^{0}_{-1}\beta$.

### ⟩⟩ Study point

Electron capture can be regarded as an equivalent to positron emission, since capture of an electron results in the same transmutation as emission of a positron.

### ◀ Stretch & challenge

Because positron emission decreases proton number relative to neutron number, positron decay happens typically in large 'proton-rich' radionuclides. Electron capture is an alternative decay mode for radioactive isotopes with insufficient energy to decay by positron emission. It therefore occurs much more often in smaller atoms than positron emission. Electron capture always competes with positron emission, however it occurs as the only type of beta decay in proton-rich nuclei when there is not enough decay energy to support positron emission.

**3 Knowledge check**

Give the mass number and symbol of the isotope formed when $^{234}$Th decays by β emission.

**4 Knowledge check**

Strontium-83 is an unstable radioactive isotope that decays by positron emission. Write an equation to show this decay.

$$^{40}_{19}\text{K} + ^{0}_{-1}\text{e}^- \longrightarrow ^{40}_{18}\text{Ar}$$

The product is one place to the left in the periodic table.

Another type of beta decay is positron emission or **β⁺ decay**. In this process a proton is converted into a neutron while releasing a positron (and an electron neutrino). The positron is a type of beta particle (β⁺).

$$^{23}_{12}\text{Mg} \longrightarrow ^{23}_{11}\text{Na} + ^{0}_{+1}\beta^+$$

The product is one place to the left in the periodic table.

# Half-life

The rate at which a radioactive isotope decays cannot be speeded up or slowed down, it is proportional to the number of radioactive atoms present. The nature of radioactive decay is shown below.

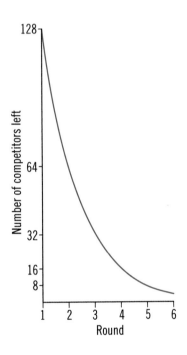

▲ Competitors at Wimbledon

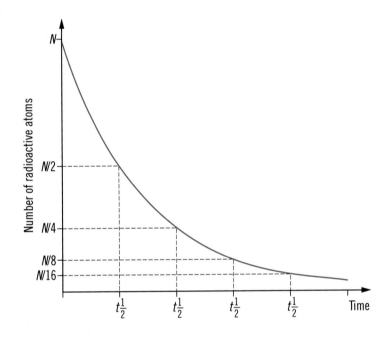

▲ Radioactive decay

The time taken for $N$ atoms to decay to $N/2$ atoms is the same as the time taken for $N/2$ atoms to decay to $N/4$ atoms and for $N/4$ atoms to decay to $N/8$ atoms. The time taken to decay to half the number of radioactive atoms is known as the **half-life**.

The process resembles a knock-out competition such as Wimbledon where one half of the competitors (atoms) disappears over each round (half-life). The number of competitors disappearing during each round (number of atoms decaying each half-life) gets smaller and smaller but is always one half of those remaining.

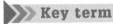
**Key term**

**Half-life** is the time taken for half the atoms in a radioisotope to decay or the time taken for the radioactivity of a radioisotope to fall to half its initial value.

There are three types of calculation involving half-life:

- Finding the time taken for the radioactivity of a sample to fall to a certain fraction of its initial value.
- Finding the mass of a radioactive isotope remaining after a certain length of time.
- Finding the half-life of a radioactive isotope.

**Worked examples**

1. The radioactive isotope $^{28}$Mg has a half-life of 21 hours.

   (a) Calculate how long it will take for the activity of the isotope to decay to ⅛ its original value.

   (b) If you started with 2.0 g of $^{28}$Mg, calculate the mass of this isotope remaining after 84 hours.

1. (a) $1 \xrightarrow{21} \dfrac{1}{2} \xrightarrow{21} \dfrac{1}{4} \xrightarrow{21} \dfrac{1}{8}$

   $21 \times 3 = 63$ hours

   (b) 84 hours = 4 half-lives

   $2.0\text{g} \xrightarrow{21} 1.0\text{g} \xrightarrow{21} 0.5\text{g} \xrightarrow{21} 0.25\text{g} \xrightarrow{21} 0.125\text{g}$

2. The radioactive isotope cobalt-60 is used in radiotherapy. Calculate its half-life if $3.6 \times 10^{-5}$ g of cobalt-60 decay to $4.5 \times 10^{-6}$ g in 15.9 years.

2. $3.6 \times 10^{-5} \longrightarrow 1.8 \times 10^{-5} \longrightarrow 9.0 \times 10^{-6} \longrightarrow 4.5 \times 10^{-6}$ is 3 half-lives.

   Half-life is 15.9/3 = 5.3 years

# Consequences for living cells

Radioactive emissions are potentially harmful. However, we all receive some radiation from the normal background radiation that occurs everywhere. Workers in industries where they are exposed to radiation from radioactive isotopes are carefully monitored to ensure that they do not receive more radiation than is allowed under internationally agreed limits.

Ionising radiation may damage the DNA of a cell. Damage to the DNA may lead to changes in the way the cell functions, which can cause mutations and the formation of cancerous cells at lower doses or cell death at higher doses.

Personal danger from ionising radiation may come from sources outside or inside the body. With a source outside the body gamma radiation is likely to be the most hazardous. However, the opposite is true for sources inside the body and if alpha particle emitting isotopes are ingested they are far more dangerous than an equivalent activity of beta emitting or gamma emitting isotopes.

# Beneficial uses of radioactivity

Although radiation from radioisotopes is harmful to health, at the same time many beneficial uses of radioactivity have been found.

## Medicine

- Cobalt-60 in radiotherapy for the treatment of cancer. The high energy of γ-radiation is used to kill cancer cells and prevent a malignant tumour from developing.
- Technetium-99m is the most commonly used medical radioisotope. It is used as a tracer, normally to label a molecule which is preferentially taken up by the tissue to be studied.

### Radio-dating

- Carbon-14 (half-life 5570 years) is used to calculate the age of plant and animal remains. All living organisms absorb carbon, which includes a small proportion of the radioactive carbon-14. When an organism dies there is no more absorption of carbon-14 and that which is already present decays. The rate of decay decreases over the years and the activity that remains can be used to calculate the age of organisms.

- Potassium-40 (half-life 1300 million years) is used to estimate the geological age of rocks. Potassium-40 can change into argon-40 by the nucleus gaining an inner electron. Measuring the ratio of potassium-40 to argon-40 in a rock gives an estimate of its age.

### Industry and analysis

- Dilution analysis. The use of isotopically labelled substances to find the mass of a substance in a mixture. This is useful when a component of a complex mixture can be isolated from the mixture in the pure state but cannot be extracted quantitatively.

- Measuring the thickness of metal strips or foil. The metal is placed between two rollers to get the right thickness. A radioactive source (a β emitter) is mounted on one side of the metal with a detector on the other. If the amount of radiation reaching the detector increases, the detector operates a mechanism for moving the rollers apart and vice versa.

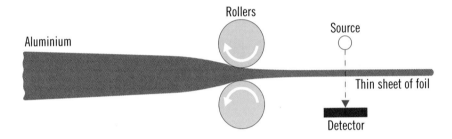

# Electronic structure

Electrons hold the key to almost the whole of chemistry since only electrons are involved in the changes that happen during chemical reactions. Electrons within atoms occupy fixed energy levels or quantum shells. Shells are numbered 1, 2, 3, 4, etc. These numbers are known as principal quantum numbers, n. The lower the value of n, the closer the shell to the nucleus and the lower the energy level.

In a quantum shell there are regions of space around the nucleus where there is a high probability of finding an electron of a given energy. These regions are called **atomic orbitals**. Orbitals of the same type are grouped together in a subshell. Each orbital can contain two electrons. Along with charge, electrons have a property called 'spin'. In order for two electrons to exist in the same orbital they must have opposite spins: this reduces the effect of repulsion. Each orbital has its own three-dimensional shape.

There is only one type of s orbital and it is spherical

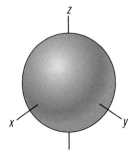

normally drawn as: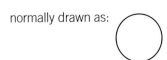

There are three different p orbitals (dumbell shaped lobes) known as the $p_x$, $p_y$ and $p_z$ orbitals. They are at right angles to each other.

These are represented as

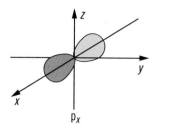 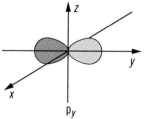 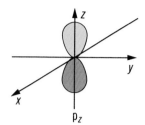

▲ Representation of p orbitals

There are five different d orbitals and seven different f orbitals.

Therefore:

- an s subshell can hold 2 electrons
- a p subshell can hold 6 electrons
- a d subshell can hold 10 electrons
- an f subshell can hold 14 electrons.

# Filling shells and orbitals with electrons

The way in which an atom's electrons are arranged in its atomic orbitals is called electronic structure or configuration. The electronic structure can be worked out using three basic rules:

1. Electrons fill atomic orbitals in order of increasing energy (Aufbau principle).

2. A maximum of two electrons can occupy any orbital each with opposite spins (Pauli exclusion principle).

3. The orbitals will first fill with one electron each with parallel spins, before a second electron is added with the paired spin (Hund's rule).

The order of filling is shown in the diagram on the right.

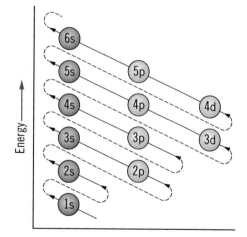

▲ The order of electron filling

An expected order is followed up to the 3p subshell, but then there is a variation, as the 4s subshell is filled before the 3d.

The most common way of representing the **electronic configuration** of an atom is to write the occupied subshells in order of increasing energy with the number of electrons following as a superscript, e.g. nitrogen has two electrons in the 1s orbital, two electrons in the 2s orbital and three electrons in the 2p subshell so it is shown as $1s^2\ 2s^2\ 2p^3$.

**◀Stretch & challenge**

The reason for the 4s orbitals filling before the 3d orbitals is due to increasingly complex influences of nuclear attractions and electron repulsions upon individual electrons.

**≫ Key term**

**Electronic configuration** is the arrangement of electrons in an atom.

For convenience sake we represent $1s^2\, 2s^2\, 2p^6\, 3s^2\, 3p^6$ as [Ar] rather than write it out each time, e.g. the electronic configuration of manganese, atomic number 25, can be written as $[Ar]\, 3d^5\, 4s^2$.

A convenient way of representing electronic configuration is using 'electrons in boxes'. Each orbital is represented as a box and the electrons are shown as arrows with their clockwise or anticlockwise spins as ↑ or ↓.

Here are the 'electrons in boxes' notation and shorter form of electronic structure for the first ten elements.

| Element | Electronic configuration | | | Electrons in boxes | | | | |
| --- | --- | --- | --- | --- | --- | --- | --- | --- |
| | | | | 1s | 2s | 2p | | |
| H | $1s^1$ | | | ↑ | | | | |
| He | $1s^2$ | | | ↑↓ | | | | |
| Li | $1s^2$ | $2s^1$ | | ↑↓ | ↑ | | | |
| Be | $1s^2$ | $2s^2$ | | ↑↓ | ↑↓ | | | |
| B | $1s^2$ | $2s^2$ | $2p^1$ | ↑↓ | ↑↓ | ↑ | | |
| C | $1s^2$ | $2s^2$ | $2p^2$ | ↑↓ | ↑↓ | ↑ | ↑ | |
| N | $1s^2$ | $2s^2$ | $2p^3$ | ↑↓ | ↑↓ | ↑ | ↑ | ↑ |
| O | $1s^2$ | $2s^2$ | $2p^4$ | ↑↓ | ↑↓ | ↑↓ | ↑ | ↑ |
| F | $1s^2$ | $2s^2$ | $2p^5$ | ↑↓ | ↑↓ | ↑↓ | ↑↓ | ↑ |
| Ne | $1s^2$ | $2s^2$ | $2p^6$ | ↑↓ | ↑↓ | ↑↓ | ↑↓ | ↑↓ |

▲ Table of electronic configuration

The electronic configuration of ions is presented in the same way as that of atoms.

Positive ions form by the loss of electrons from the highest energy orbitals so these ions have fewer electrons than the parent atom.

Negative ions form by adding electrons to the highest energy orbitals so these ions have more electrons than the parent atom,

e.g. Na $1s^2\, 2s^2\, 2p^6\, 3s^1$        $Na^+$ $1s^2\, 2s^2\, 2p^6$

   Cl $1s^2\, 2s^2\, 2p^6\, 3s^2\, 3p^5$        $Cl^-$ $1s^2\, 2s^2\, 2p^6\, 3s^2\, 3p^6$

# Ionisation energies

The process of removing electrons from an atom is called ionisation. The energy needed to remove each successive electron from an atom is called the first, second, third, etc., ionisation energy.

The process for the **first ionisation energy** (IE) of an element is summarised in the equation:

$$X(g) \longrightarrow X^+(g) + e^-$$

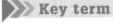

### Key term

**The first ionisation energy** of an element is the energy required to remove one electron from each atom in one mole of its gaseous atoms.

Electrons are held in their shells by their attraction to the positive nucleus, therefore the greater the attraction, the greater the ionisation energy. This attraction depends on three factors:

- The size of the positive nuclear charge – the greater the nuclear charge, the greater the attractive force on the outer electron and the greater the ionisation energy.

- The distance of the outer electron from the nucleus – the force of attraction between the nucleus and the outer electron decreases as the distance between them increases. The further an electron is from the nucleus, the lower the ionisation energy.

- The **shielding effect** by electrons in filled inner shells – all electrons repel each other since they are negatively charged. Electrons in the filled inner shells repel electrons in the outer shell and reduce the effect of the positive nuclear charge. The more filled inner shells or subshells there are, the smaller the attractive force on the outer electron and the lower the ionisation energy.

Evidence for shells and subshells can be seen from a plot of first ionisation energies against the elements (or atomic number). Such a plot is shown below for the first twenty elements.

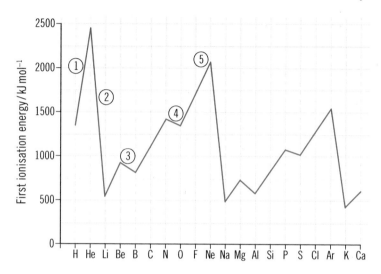

▲ Plot of IE against elements

The most significant features of the plot are:

- The 'peaks' are occupied by elements of Group 0.
- The 'troughs' are occupied by elements of Group 1.
- There is a general increase in ionisation energy across a period, although this increase is not uniform.
- There is a decrease in ionisation energy going down a group.

Looking at the plot in detail (remember the three main factors that affect ionisation energy):

1. He > H since helium has a greater nuclear charge in the same subshell so little extra shielding.

2. He > Li since lithium's outer electron is in a new shell which has increased shielding and is further from the nucleus.

3. Be > B since boron's outer electron is in a new subshell of slightly higher energy level and is partly shielded by the 2s electrons.

4. N > O since the electron–electron repulsion between the two paired electrons in one p orbital in oxygen makes one of the electrons easier to remove. Nitrogen does not contain paired electrons in its p orbital.

5. He > Ne since neon's outer electron has increased shielding from inner electrons and is further from the nucleus.

>> **Key term**

**Shielding effect** is the repulsion between electrons in different shells. Inner shell electrons repel outer shell electrons.

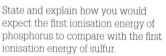

**⟨ Link ⟩**

Trends across periods page 65

**Knowledge check  9**

State and explain how you would expect the first ionisation energy of phosphorus to compare with the first ionisation energy of sulfur.

>> **Study point**

If the conditions for ionisation energy are 298 K and 1 atm then the process is known as the standard ionisation energy.

All ionisation energies are positive since it always requires energy to remove an electron.

# Successive ionisation energies

Further evidence for shells and subshells comes from the **successive ionisation energies** needed to remove all the electrons from an atom.

An element has as many ionisation energies as it has electrons. Sodium has eleven electrons and so has eleven successive ionisation energies.

For example, the third ionisation energy is a measure of how easily a 2+ ion loses an electron to form a 3+ ion. An equation to represent the third ionisation energy of sodium is:

$$Na^{2+}(g) \longrightarrow Na^{3+}(g) + e^-$$

Successive ionisation energies always increase because:

- There is a greater effective nuclear charge as the same number of protons are holding fewer and fewer electrons.
- As each electron is removed there is less electron–electron repulsion and each shell will be drawn in slightly closer to the nucleus.
- As the distance of each electron from the nucleus decreases, the nuclear attraction increases.

As the ionisation energies are so large we must use logarithms to base 10 ($\log_{10}$) to make the numbers fit on a reasonable scale. Remember, electrons are removed in order, starting with the furthest from the nucleus.

The graph below shows the successive ionisation energies of sodium.

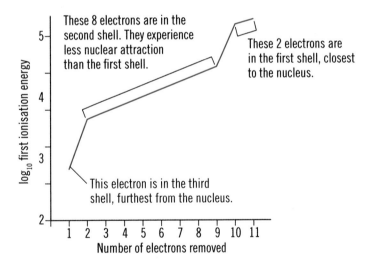

▲ Graph of sodium's IE

For sodium there is one electron on its own which is easiest to remove. Then there are eight more electrons which become successively more difficult to remove. Finally there are two electrons which are the most difficult to remove.

Notice the large increases in ionisation energy as the 2nd and 10th electrons are removed. If the electrons were all in the same shell, there would be no large rise or jump.

# Emission and absorption spectra

## Light and electromagnetic radiation

Light is a form of electromagnetic radiation. Electromagnetic radiation is energy travelling as waves. A wave is described by its frequency ($f$) and its wavelength ($\lambda$).

The frequency and wavelength of light are related by the equation:

$c = f\lambda$   ($c$ is the speed of light)

The frequency of electromagnetic radiation and energy ($E$) are connected by the equation:

$E = hf$   ($h$ is Planck's constant)

Therefore, $f \propto E$, and if frequency increases, energy increases.

$f \propto 1/\lambda$ and if frequency increases, wavelength decreases.

The whole range of frequencies of electromagnetic radiation is called the electromagnetic spectrum.

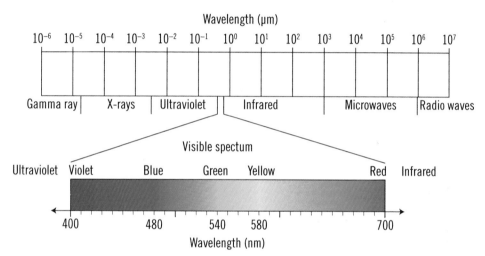

▲ The electromagnetic spectrum

In this unit we are only concerned with the infrared, visible and ultraviolet regions.

It is important to note that both energy and frequency increase going from the infrared through visible to the ultraviolet region. Thus, blue light is of a higher energy than red light. As frequency gets larger and therefore wavelength decreases, blue light must have a shorter wavelength than red light.

## Absorption spectra

Light of all visible wavelengths is called white light. All atoms and molecules absorb light of certain wavelengths. Therefore, when white light is passed through the vapour of an element, certain wavelengths will be absorbed by the atoms and removed from the light. Looking through a spectrometer, black lines appear in the spectrum where light of some wavelengths has been absorbed. The wavelengths of these lines correspond to the energy taken in by the atoms to promote electrons from lower to higher energy levels.

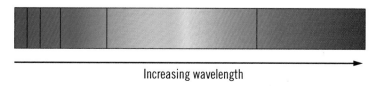

Increasing wavelength

# Emission spectra

When atoms are given energy by heating or by an electrical field, electrons are excited and the additional energy promotes them from a lower energy level to a higher one. When the source of energy is removed and the electrons leave the excited state, they fall from the higher energy level to a lower energy level and the energy lost is released as a photon (a quantum of light energy) with a specific frequency. The observed spectrum consists of a number of coloured lines on a black background.

>> **Study point**

A spectrometer is an instrument that separates light into its constituent wavelengths.

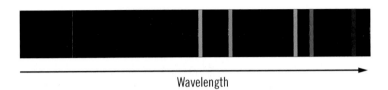

Wavelength

The fact that only certain colours appear in an atom's emission spectrum means that only photons having certain energies are emitted by the atom.

If the electron energy levels were not quantised but could have any value, a continuous spectrum rather than a line spectrum would result.

**12 Knowledge check**

Give two differences between absorption and emission spectra.

# The hydrogen spectrum

An atom of hydrogen has only one electron so it gives the simplest emission spectrum. The atomic spectrum of hydrogen consists of separate series of lines mainly in the ultraviolet, visible and infrared regions of the electromagnetic spectrum. There are six series, each named after their discoverer. Only one series, the Balmer series, is in the visible region of the spectrum.

**13 Knowledge check**

Which letter represents the transition that causes a line of the lowest frequency in the emission spectrum of atomic hydrogen?

A  Second energy level to first energy level.

B  Third energy level to first energy level.

C  Third energy level to second energy level.

D  Fourth energy level to second energy level.

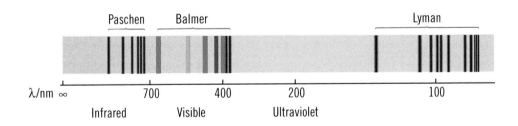

▲ Diagram showing part of the emission spectrum of atomic hydrogen.

When an atom is excited by absorbing energy, an electron jumps up to a higher energy level. As the electron falls back down to a lower energy level it emits energy in the form of electromagnetic radiation. The emitted energy can be seen as a line in the spectrum because the energy of the emitted radiation is equal to the difference between the two energy levels, $\Delta E$, in this electronic transition, i.e. it is a fixed quantity or quantum.

Since $\Delta E = hf$, electronic transitions between different energy levels result in emission of radiation of different frequencies and therefore produce different lines in the spectrum.

>> **Study point**

Electronic transition is when an electron moves from one energy level to another.

As the frequency increases, the lines get closer together because the energy difference between the shells decreases. Each line in the Lyman series (ultraviolet region) is due to electrons returning to the first shell or $n = 1$ energy level, while the Balmer series (visible region) is due to electrons returning to the second shell or $n = 2$ energy level.

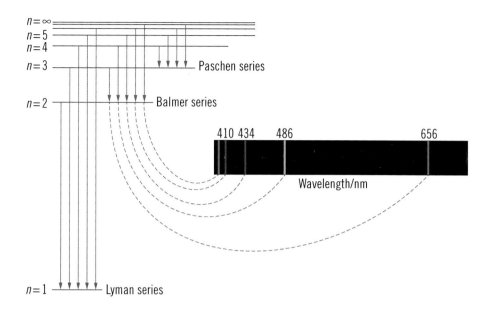

>> **Study point**

The various series in the infrared region are caused by electrons returning to the energy levels $n = 3$ (Paschen), $n = 4$ (Brackett), $n = 5$ (Pfund) and $n = 6$ (Humphreys).

You do not need to know the names of these series.

# Ionisation energy of the hydrogen atom

The spectral lines become closer and closer together as the frequency of the radiation increases until they converge to a limit. The **convergence limit** corresponds to the point at which the energy of an electron is no longer quantised. At that point the nucleus has lost all influence over the electron; the atom has become ionised.

For the Lyman series, $n = 1$, the convergence limit represents the ionisation of the hydrogen atom.

Measuring the convergent frequency (difference from $n = 1$ to $n = \infty$) allows the ionisation energy to be calculated using $\Delta E = hf$.

The value of $\Delta E$ is multiplied by Avogadro's constant to give the first ionisation energy for a mole of atoms.

>> **Key term**

The **convergence limit** is when the spectral lines become so close together they have a continuous band of radiation and separate lines cannot be distinguished.

<< **Exam tip**

The IE of a hydrogen atom can be shown on its electron energy level diagram by drawing an arrow upwards from the $n = 1$ to the $n = \infty$ level.

<< **Maths tip**

Wavelength has to be in **metres**. Don't forget to change 'nm' to 'm' by multiplying by $10^{-9}$.

**Worked example**

The value of the wavelength at the start of the continuum in the hydrogen emission spectrum is 92 nm. Calculate the first ionisation energy of hydrogen.

(Assume that $c = 3.00 \times 10^8 \, \text{m s}^{-1}$, $h = 6.63 \times 10^{-34} \, \text{Js}$ and $N_A = 6.02 \times 10^{23} \, \text{mol}^{-1}$)

Ionisation energy $= N_A \Delta E$   ($N_A$ = Avogadro's constant)

But $\Delta E = hf$ and $f = c/\lambda$   ($h$ = Planck's constant, $c$ = speed of light)

Therefore ionisation energy $= N_A hc/\lambda$

$$= 6.02 \times 10^{23} \times 6.63 \times 10^{-34} \times \frac{3.00 \times 10^8}{(92 \times 10^{-9})}$$

$$= 1\,301\,498 \, \text{J mol}^{-1}$$

$$= 1301 \, \text{kJ mol}^{-1}$$

*Continued* ▶

**14** **Knowledge check**

The value of the frequency at the start of the continuum in the sodium emission spectrum is $1.24 \times 10^{15}$ Hz. Calculate the first ionisation energy of sodium in kJ mol$^{-1}$.

If you don't want to combine all the equations together and do one large calculation, you can use each equation separately and perform three small calculations.

$$\text{Frequency} = \frac{c}{\lambda} = \frac{3.00 \times 10^8}{92 \times 10^{-9}} = 3.26 \times 10^{15} \text{ (Hz)}$$

$$\text{Energy} = hf = 6.63 \times 10^{-34} \times 3.26 \times 10^{15} = 2.16 \times 10^{-18} \text{ (J)}$$

$$\text{Ionisation energy} = N_A E = 6.02 \times 10^{23} \times 2.16 \times 10^{-18} = 1301498 \text{ (J mol}^{-1})$$

$$= 1301 \text{ kJ mol}^{-1}$$

# Test yourself

1. (a) Complete the table below, which relates to the particles contained in an atom. [3]

| Particle | Relative mass | Relative charge |
|---|---|---|
| Proton | | |
| | 1 | |
| | | |

(b) An element, X, has an atomic number of 47 and forms an ion X$^+$.

State the number of protons and electrons in this **ion**. [1]

(c) The symbols $^{23}_{11}$Na, $^{63}_{29}$Cu and $^{65}_{29}$Cu and, represent sodium atoms and copper atoms respectively.

Use these symbols to explain the meaning of the terms:

(i) Mass number [2]

(ii) Isotope [2]

(d) Name the species which, on gaining an electron, has a full set of 2p orbitals. [1]

(e) Show the electronic configuration of a titanium atom. [1]

2. (a) One of the unstable isotopes of platinum, $^{199}$Pt, decays by β-emission.

(i) Write an equation to show this decay. [1]

(ii) State what happens in the nucleus of an atom when a β-particle forms. [1]

(iii) 10.0 mg of $^{199}$Pt decays to 1.25 mg in one hour and 33 minutes. Calculate the half-life of this isotope. [1]

(b) $^{23}$Mg decays by positron emission which is another type of β-decay.

Give the mass number and symbol of the species produced when 1 atom of $^{23}$Mg decays. [1]

(c) Describe why radioactivity is dangerous to living cells. Explain which type of radiation causes most damage to cells. [4]

3.  (a)  The first four ionisation energies for an element, X, are shown in the table below.

| Ionisation energy / kJ mol$^{-1}$ | | | |
|---|---|---|---|
| 1st | 2nd | 3rd | 4th |
| 1000 | 2260 | 3390 | 4540 |

State what you can deduce about the group to which the element belongs. Explain your answer. [2]

(b)  Nitrogen has a first ionisation energy of 1400 kJ mol$^{-1}$. State and explain how you would expect the first ionisation energy of nitrogen to compare with the first ionisation energy of:

(i)   Carbon [2]

(ii)  Oxygen [2]

(iii) Phosphorus [2]

(c)  Successive ionisation energies are a measure of the energy needed to remove each electron in turn until all the electrons are removed from an atom.

(i)   Write an equation to represent the third ionisation energy of nitrogen. [1]

(ii)  Explain how successive ionisation energies give evidence for shells and subshells in an atom. [2]

4.  Describe the main features of the atomic absorption spectrum of hydrogen. Explain how these features arise and how their interpretation provides evidence for energy levels in the atom. [5]

5.  (a)  The atomic emission spectrum of hydrogen consists of separate series of lines caused by electrons being excited and falling back down to a lower energy level.

When an electron returns from the fifth shell to the second shell, the energy of the radiation emitted is $4.58 \times 10^{-22}$ kJ. Calculate the wavelength, in nm, at which the line can be seen. [4]

(b)  State whether the energy of radiation emitted at a wavelength of 656 nm would be greater or less than $4.58 \times 10^{-22}$ kJ. Explain your answer. [1]

# Chemical calculations

Chemists are interested in how atoms react together. They want to know about the qualitative and quantitative aspects of chemical reactions. Very often chemists want to measure out exact quantities of substances that will react together – especially in industry where adding too much of a reagent will result in an unnecessary cost or may contaminate the product. What is really useful is to be able to work out these quantities on the basis of the number of atoms (or molecules) of substances that react together. This topic shows how this can be done by using the concept of the mole and balanced chemical equations.

## Topic contents

You should be able to demonstrate and apply knowledge and understanding of:

- The various relative mass terms (atomic, isotopic, formula, molecular).
- The principles of the mass spectrometer and its use in determining relative atomic mass and relative abundance of isotopes.
- Simple mass spectra, for example that of chlorine gas.
- How empirical and molecular formulae can be determined from given data.
- The relationship between the Avogadro constant, the mole and molar mass.
- The relationship between grams and moles.
- The concept of concentration and its expression in terms of grams or moles per unit volume (including solubility).
- Molar volume and correction due to changes in temperature and pressure.
- The ideal gas equation ($pV = nRT$).
- The concept of stoichiometry and its use in calculating reacting quantities, including in acid–base titrations.
- The concepts of atom economy and percentage yield.
- How to estimate the percentage error in a measurement and use this to express numeric answers to a sensible number of significant figures.

## Maths skills ⟫

- Use appropriate units in calculations.
- Substitute numerical values into algebraic equations.
- Change the subject of an equation.
- Use standard and ordinary form, decimal places and significant figures.
- Select values and find arithmetic means.
- Identify uncertainties in measurements.
- Use ratios, fractions and percentages.

# Relative mass terms

## Masses of atoms

The masses of individual atoms are too small to be used in calculations in chemical reactions, so instead the mass of an atom is expressed relative to a chosen standard atomic mass. The carbon-12 isotope is taken as the standard of reference because relative atomic masses are determined by mass spectrometry and volatile carbon compounds are widely used in mass spectrometry.

Most elements exist naturally as two or more different isotopes. The mass of an element therefore depends on the relative abundance of all the isotopes present in the sample. In order to overcome this, chemists use an average mass of all the atoms and this is called the **relative atomic mass, $A_r$**.

Relative atomic mass has no units since it is one mass compared with another mass.

If we refer to the mass of a particular isotope then the term **relative isotopic mass** is used.

## Masses of compounds

Since the formula of a compound shows the ratio in which the atoms combine, the idea of relative atomic mass can be extended to compounds and the term **relative formula mass, $M_r$**, is used.

For example, the relative formula mass of copper(II) sulfate, $CuSO_4$ is:

$(1 \times 63.5) + (1 \times 32) + (4 \times 16) = 159.5$

# The mass spectrometer

When a mass spectrometer is used to find the relative atomic mass of an element it measures two things:

- The mass of each different isotope of the element.
- The relative abundance of each isotope of the element.

The diagram below shows how a mass spectrometer works.

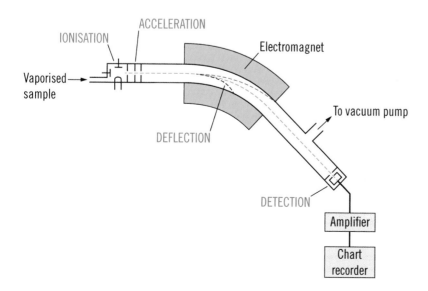

> **Key terms**

**Relative atomic mass** is the average mass of one atom of the element relative to one-twelfth the mass of one atom of carbon-12.

**Relative isotopic mass** is the mass of an atom of an isotope relative to one-twelfth the mass of an atom of carbon-12.

**Relative formula mass** is the sum of the relative atomic masses of all atoms present in its formula.

> **Study point**

Relative formula mass used to be called relative molecular mass, but relative molecular mass really refers to compounds containing only molecules.

> **Knowledge check**

Calculate the relative formula mass for
(a) $Ca(OH)_2$
(b) $Na_2S_2O_3.5H_2O$

### ◄Stretch & challenge

During the detecting stage, when an ion hits the metal box, its charge is neutralised by an electron jumping from the metal to the ion. That leaves a space amongst the electrons in the metal, and the electrons in the wire shuffle along to fill it.

A flow of electrons in the wire is detected as an electric current which can be amplified and recorded. The more ions arriving, the greater the current.

### ▲ 2 Knowledge check

State how positive ions are produced in a mass spectrometer.

### ▲ 3 Knowledge check

A mass spectrum of a sample of hydrogen showed that it contained $^1H$ 99.20% and $^2H$ 0.8000%.

Calculate the relative atomic mass of the hydrogen sample, giving your answer to **four** significant figures.

---

There are four main steps:

**1. Ionisation** The vaporised sample passes into the ionisation chamber.

The particles in the sample (atoms or molecules) are bombarded with a stream of electrons, and some of the collisions are sufficiently energetic to knock one or more electrons out of the sample particles to make positive ions.

Most of the positive ions formed will carry a charge of +1 because it is much more difficult to remove further electrons from an already positive ion.

**2. Acceleration** An electric field accelerates the positive ions to high speed.

**3. Deflection** Different ions are deflected by the magnetic field by different amounts. The amount of deflection depends on:

- The mass of the ion. Lighter ions are deflected more than heavier ones.
- The charge on the ion. Ions with two (or more) positive charges are deflected more than ones with only one positive charge.

These two factors are combined into the mass/charge ratio $m/z$.

**(Unless otherwise stated, mass spectra given will only involve 1+ ions, so the mass/charge ratio will be the same as the mass of the ion.)**

**4. Detection** The beam of ions passing through the machine is detected electrically. Only ions with the correct mass/charge ratio make it all the way through the machine to the ion detector. (The other ions collide with the walls where they will pick up electrons and be neutralised. Eventually, they get removed from the mass spectrometer by the vacuum pump.) The signal is then amplified and recorded.

For the ions with an incorrect mass/charge ratio to reach the detector, the strength of the magnetic field can be varied. For example, in the diagram on page 33, the beam of ions with the black dotted line is the most deflected. For the ions to reach the detector, they need to be deflected less so a smaller magnetic field would be necessary.

It's important that the ions produced in the ionisation chamber have a free run through the machine without hitting air molecules so a **vacuum** is needed inside the apparatus.

## Calculating relative atomic mass

As stated, data from mass spectrometry can be used to calculate relative atomic mass, e.g. on the right is the mass spectrum of lead

The relative atomic mass is a weighted average of the masses of all the atoms in the isotopic mixture, therefore:

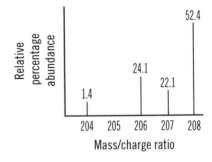

relative atomic mass =

$$\frac{(1.40 \times 204) + (24.1 \times 206) + (22.1 \times 207) + (52.4 \times 208)}{100}$$

$$= 207.2$$

Other uses of mass spectrometry include:

- Identifying unknown compounds, e.g. testing athletes for prohibited drugs.
- Identifying trace compounds in forensic science.
- Analysing molecules in space.

## Interpretation of mass spectra

When a vaporised compound passes through a mass spectrometer, an electron is knocked off a molecule to form a positive ion. This ion is called the **molecular ion** and its mass gives the relative formula mass of the compound.

The molecular ions are energetically unstable, and some of them will break up into smaller pieces or fragments.

All sorts of fragmentations of the original molecular ion are possible – which means that you will get a wide range of lines in the mass spectrum.

## The mass spectrum of chlorine

Chlorine is made up of two isotopes $^{35}$Cl and $^{37}$Cl. Chlorine gas consists of molecules, not individual atoms, but the mass spectrum of chlorine is:

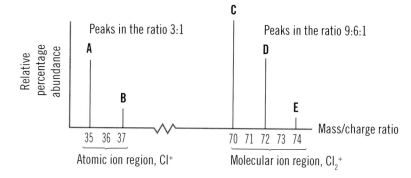

When chlorine is passed into the ionisation chamber, an electron is knocked off the molecule to give a molecular ion, $Cl_2^+$. These ions won't be particularly stable, and some will fall apart to give a chlorine atom and a $Cl^+$ ion. (This is known as **fragmentation**.)

So peak A is caused by $^{35}Cl^+$ and peak B by $^{37}Cl^+$.

As the $^{35}$Cl isotope is three times more common than the $^{37}$Cl isotope, the heights of the peaks are in the ratio of 3:1.

In the molecular ion region think about the possible combinations of $^{35}$Cl and $^{37}$Cl atoms in a $Cl_2^+$ ion. Both atoms could be $^{35}$Cl, both atoms could be $^{37}$Cl, or you could have one of each sort.

So peak C (*m/z* 70) is due to $(^{35}Cl\text{—}^{35}Cl)^+$

Peak D (*m/z* 72) is due to $(^{35}Cl\text{—}^{37}Cl)^+$ or $(^{37}Cl\text{—}^{35}Cl)^+$

Peak E (*m/z* 74) is due to $(^{37}Cl\text{—}^{37}Cl)^+$

Since the probability of an atom being $^{35}$Cl is $\frac{3}{4}$ and that of being $^{37}$Cl is $\frac{1}{4}$, then

| molecule | $^{35}Cl\text{—}^{35}Cl$ | $^{35}Cl\text{—}^{37}Cl$ or $^{37}Cl\text{—}^{35}Cl$ | $^{37}Cl\text{—}^{37}Cl$ |
|---|---|---|---|
| probability | $\frac{3}{4} \times \frac{3}{4}$ | $\frac{3}{4} \times \frac{1}{4}$ or $\frac{1}{4} \times \frac{3}{4}$ | $\frac{1}{4} \times \frac{1}{4}$ |
| | $\frac{9}{16}$ | $\frac{6}{16}$ | $\frac{1}{16}$ |

and ratio of peaks C : D : E is 9 : 6 : 1

The molecular ion region of mass spectra can give information about the isotopes in the molecule.

Bromine has two isotopes. The molecular ion region of its mass spectrum is shown on the right.

 **Link**

Mass spectrometry page 177

**Study point**

For the chlorine spectrum you can't make any predictions about the relative heights of the lines at *m/z* 35/37 compared with those at 70/72/74. That depends on what proportion of the molecular ions break up into fragments.

**Knowledge check** 4

In the mass spectrum of hydrogen, explain why peaks due to hydrogen atoms are present although hydrogen gas contains only $H_2$ molecules.

**Key terms**

**Molecular ion** is a positive ion formed in a mass spectrometer from the whole molecule.

**Fragmentation** is splitting of molecules, in a mass spectrometer, into smaller parts.

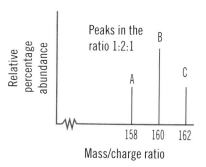

Since bromine only has two isotopes:

peak A (*m/z* 158) must be due to ($^{79}$Br—$^{79}$Br)$^+$ and

peak C (*m/z* 162) must be due to ($^{81}$Br—$^{81}$Br)$^+$.

Therefore peak B (m/z 160) is due to ($^{79}$Br—$^{81}$Br)$^+$ or ($^{81}$Br—$^{79}$Br)$^+$.

Since the ratio of peaks A : B : C is almost 1 : 2 : 1 it follows that natural bromine consists of a nearly 50:50 mixture of $^{79}$Br and $^{81}$Br.

| molecule | $^{79}$Br—$^{79}$Br | $^{79}$Br—$^{81}$Br or $^{81}$Br—$^{79}$Br | $^{81}$Br—$^{81}$Br |
|---|---|---|---|
| probability | $\frac{1}{2} \times \frac{1}{2}$ | $\frac{1}{2} \times \frac{1}{2}$ or $\frac{1}{2} \times \frac{1}{2}$ | $\frac{1}{2} \times \frac{1}{2}$ |
| | $\frac{1}{4}$ | $\frac{2}{4}$ | $\frac{1}{4}$ |

# Amount of substance

In chemical reactions, for all the reactants to change into products, the correct quantity of each reactant must be used. Since the atoms that make up the reactants rearrange to form the products, it would be really useful to be able to work out these quantities on the basis of the number of atoms that react.

If you pay coins into a bank, the clerk won't count them individually – he knows the mass of a certain number of coins, e.g. £10 worth of 50p, so he will weigh a bag of coins to confirm that the correct number of coins is present.

Similarly, atoms are too small to be counted individually, so they are counted by weighing a collection of them where the mass of a particular fixed number of atoms is known.

Since carbon-12 is the standard chosen for relative atomic mass, the number of atoms in exactly 12g of carbon-12 is chosen as the standard and is known as the **mole**.

The number of atoms in a mole is very large. Using a mass spectrometer, the mass of a single carbon-12 atom has been found to be $1.993 \times 10^{-23}$ g therefore

$$\text{the number of atoms per mole} = \frac{\text{mass per mole of } ^{12}\text{C}}{\text{mass of one atom of } ^{12}\text{C}}$$

$$= \frac{12\,\text{g mol}^{-1}}{1.993 \times 10^{-23}\,\text{g}}$$

$$= 6.02 \times 10^{23}\,\text{mol}^{-1}$$

This is called the **Avogadro constant**, $N_A$, after the nineteenth-century Italian chemist Amadeo Avogadro.

The mass of one mole of a substance is called the **molar mass**, **M**. It has the same numerical value as $A_r$ or $M_r$ but has the unit g mol$^{-1}$.

The amount of substance in moles (*n*), the mass (*m*) and molar mass (*M*) are linked by the equation:

$$\text{amount in moles } (n) = \frac{\text{mass } (m)}{\text{molar mass } (M)}$$

The expression for amount in moles may be rearranged in two ways to find the mass of a sample or molar mass of a substance.

**Worked examples**

What is the amount of sodium present in 0.23 g of sodium?

$A_r$ of sodium = 23.0

$$\text{Molar mass of sodium} = 23.0 \, \text{g mol}^{-1}$$

$$\text{Amount of sodium} = \frac{m}{M} = \frac{0.23 \, \text{g}}{23.0 \, \text{g mol}^{-1}} = 0.01 \, \text{mol}$$

If you need 0.05 mol of sodium hydroxide, what mass of the substance do you have to weigh out?

$$\text{Molar mass of NaOH} = 40 \, \text{g mol}^{-1}$$

$$\text{Mass of sample} = n \times M$$

$$= 0.05 \times 40$$

$$= 2 \, \text{g}$$

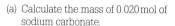

## Knowledge check 6

(a) Calculate the mass of 0.020 mol of sodium carbonate.

(b) Calculate the amount (in moles) of 1.36 g calcium carbonate.

# Calculating reacting masses

In a chemical reaction, reactants change into products. If we are given the mass of reactants, we can find out the mass of products formed as long as we have a balanced chemical equation for the reaction.

An equation tells us not only what substances react together but also what amounts of substances, in moles, react together. The ratio between amounts in moles of reactants and products are called the **stoichiometric** ratios (mole ratios).

E.g. the equation for the burning of magnesium in oxygen to produce magnesium oxide

$$2Mg + O_2 \longrightarrow 2MgO$$

tells us that 2 moles of magnesium react with 1 mole of oxygen to produce 2 moles of magnesium oxide.

What mass of magnesium oxide forms if we burn 1.215 g of magnesium in an excess of oxygen?

To calculate this, the following route is used:

**Step 1** Change the mass of Mg into amount of moles (divide by molar mass).

**Step 2** Use the balanced equation to state the mole ratio of Mg : MgO, hence deduce the moles of MgO.

**Step 3** Change the amount in moles of MgO to mass (multiply by molar mass).

**Step 1** Amount in moles of Mg $= \dfrac{1.215}{24.3} = 0.050 \, \text{mol}$

**Step 2** The mole ratio from the equation is 2Mg : 2MgO i.e. 1 : 1

therefore 0.050 mol Mg gives 0.050 mol MgO

**Step 3** Molar mass of MgO = 24.3 + 16 = 40.3

$$\text{Mass MgO} = 0.05 \times 40.3$$

$$= 2.015 \, \text{g}$$

## Key term

**Stoichiometry** is the molar relationship between the amounts of reactants and products in a chemical reaction.

## Study point

Excess oxygen implies that there is more than enough oxygen present to react with all the magnesium burnt in it. When the reaction is complete, all the magnesium has reacted and some oxygen remains.

## Knowledge check 7

In a laboratory, iron can be produced by reducing iron(III) oxide with aluminium.

$$Fe_2O_3 + 2Al \longrightarrow 2Fe + Al_2O_3$$

What mass of iron(III) oxide is needed to produce 1.20 g of iron?

## Stretch & challenge

When iron ore (iron(III) oxide) is reduced by carbon monoxide in a blast furnace, iron is produced. Calculate how much iron is produced from 1 kg of iron ore.

# Empirical and molecular formulae

We can use calculations involving masses of the elements that combine together to find the formulae of compounds.

**Empirical formula** is the simplest formula showing the simplest whole number ratio of the number of atoms of each element present.

**Molecular formula** shows the actual number of atoms of each element present in the molecule. It is a simple multiple of the empirical formula. Usually the relative formula mass is needed to determine the molecular formula.

Empirical formulae can be calculated from known masses or percentage composition data. There are three steps in the calculation:

**Step 1** Find the amount in moles of each element present (divide by the molar mass).

**Step 2** Find the ratio of the number of atoms present (divide by the smallest value in step 1).

**Step 3** Convert these numbers into whole numbers (atoms combine together in whole number ratios).

## Maths tip

After you've divided the percentages or masses by the relevant atomic masses, do not truncate the answers, e.g. 1.2 to 1.

After you've divided by the smallest number only round off to a whole number if the answer is very close to that number, e.g. round off 2.03 to 2 but not 2.25 to 2.

**8 Knowledge check**

Find the empirical formula of the compound formed when 1.172 g iron forms 3.409 g of iron chloride.

---

**Worked example**

A compound of carbon, hydrogen and oxygen has a relative molecular mass of 60. The percentage composition by mass is C 40.0%; H 6.70%; O 53.3%.

What is (a) the empirical formula and (b) the molecular formula?

(a)

|  | C | : | H | : | O |
|---|---|---|---|---|---|
| Molar ratio of atoms | $\frac{40}{12}$ | | $\frac{6.7}{1.01}$ | | $\frac{53.3}{16}$ |
| | 3.33 | | 6.63 | | 3.33 |
| Divide by smallest number | 1 | | 1.99 | | 1 |

Empirical formula is $CH_2O$

(b) Mass of empirical formula $= 12 + 2.02 + 16 = 30.02$

Number of $CH_2O$ units in a molecule $= \frac{60}{30.02} = 2$

Molecular formula is $C_2H_4O_2$

---

Some salts have water molecules incorporated into their structure. These are known as hydrated salts and the water is known as water of crystallisation. If we know the mass of the anhydrous salt and the mass of the water in the hydrated salt we can calculate the number of moles of water in the hydrated salt.

---

**Worked example**

Sodium carbonate can form a hydrate, $Na_2CO_3.xH_2O$. When 4.64 g of this hydrate were heated, 2.12 g of the anhydrous salt, $Na_2CO_3$, remained.

What is the value of $x$?

Mass of water in the hydrate $= 4.64 - 2.12 = 2.52$ g

Moles $Na_2CO_3 = \frac{2.12}{106} = 0.020$

Moles $H_2O = \frac{2.52}{18.02} = 0.140$

|  | $H_2O$ | : | $Na_2CO_3$ |
|---|---|---|---|
| Mole ratio of | 0.140 | : | 0.020 |
| Divide by smaller number | 7 | : | 1 |

Value of $x = 7$ and formula is $Na_2CO_3.7H_2O$

---

# Volumes of gases

For reactions involving gases, it is more usual to consider the volumes of reactants and products rather than their masses. This provides a way of calculating the amount of gas present. At standard temperature and pressure (stp), 0°C and 1 atm, one mole of any gas occupies $24.4\,dm^3$. This is known as the gas **molar volume**, $v_m$.

At room temperature and pressure (rtp), 25 °C and 1 atm, one mole of gas occupies $24.5\,dm^3$.

The steps in calculations involving molar volume are similar to the ones for reacting masses.

For example, what volume of hydrogen is produced, at 25 °C and 1 atm, where 3.00 g of zinc reacts with excess hydrochloric acid?

(1 mole of hydrogen occupies $24.5\,dm^3$ at 25 °C and 1 atm)

$$Zn + 2HCl \longrightarrow ZnCl_2 + H_2$$

**Step 1**  Change the mass of zinc into moles

Amount of moles of zinc $= \dfrac{3.00}{65.4} = 0.0459$

**Step 2**  The mole ratio from the equation is $1Zn : 1H_2$

therefore 0.0459 mol Zn gives 0.0459 mol $H_2$

**Step 3**  Change the moles into volume of gas

Volume of hydrogen $= 0.0459 \times 24.5 = 1.12\,dm^3$

# The ideal gas equation

In the real world, most reactions are not carried out at room temperature and pressure. The number of moles in a certain volume of gas at any temperature and pressure can be calculated using the ideal gas equation:

$pV = nRT$

where $p$ is the pressure measured in Pa (pascals)

$V$ is the volume measured in $m^3$

$n$ is the number of moles

$R$ is the gas constant and has the value of $8.31\ JK^{-1}mol^{-1}$

$T$ is the temperature measured in K (kelvins)

## Key term

**Molar volume, $v_m$,** is the volume per mole of a gas. (In the exam, it will be given on the data sheet.)

## Study point

To calculate amount of moles, $n$, of a gas from volume, use the equation
$v = n \times v_m$

Remember that the unit for $v_m$ is $dm^3$.

## Exam tip

Only use molar volume if the temperature is 0 °C or 25 °C and the pressure is 1 atm.

## Knowledge check

Calculate the volume of carbon dioxide produced when 3.40 g of calcium carbonate is heated and decomposes according to the equation

$CaCO_3(s) \longrightarrow CaO(s) + CO_2(g)$.

(Assume that 1 mol of gas occupies $24.0\,dm^3$ under the experimental conditions.)

## Maths tip

It is vital to use the correct units when using the ideal gas equation, i.e. Pa, $m^3$ and K.

For pressure, 1atm = $1.01 \times 10^5$ Pa

To change kPa to Pa multiply by 1000.

Volumes are normally measured in $cm^3$ or $dm^3$.

Since $1 m^3$ = 1000 $dm^3$
and 1 $dm^3$ = 1000 $cm^3$

to change from $m^3$ to $cm^3$ multiply by $10^6$

to change from $cm^3$ to $m^3$ divide by $10^6$ or multiply by $10^{-6}$.

Temperatures are often measured in °C.

To change from °C to K add 273

so 25 °C = 298 K.

**10 Knowledge check**

State the units that must be used for pressure, volume and temperature in the ideal gas equation.

**11 Knowledge check**

Calculate the volume, in dm³, occupied by 0.10 mol of carbon dioxide at $1.01 \times 10^5$ Pa and 110 °C.

(Gas constant $R = 8.31 \, J \, K^{-1} \, mol^{-1}$)

**▶▶ Study point**

In calculations involving $\dfrac{p_1 V_1}{T_1} = \dfrac{p_2 V_2}{T_2}$ remember to change temperature to kelvin.

**Maths tip ▶▶**

To measure the volume of a gas:

at a different temperature but same pressure use $p_1 V_1 = p_2 V_2$.

at a different pressure but same temperature use

$$\dfrac{V_1}{T_1} = \dfrac{V_2}{T_2}$$

**12 Knowledge check**

Calculate the volume of a gas at stp if it occupies 200 cm³ at 50 °C and 2 atm.

(Conditions at stp are 273 K and 1 atm)

**Worked example**

A sample of sulfur dioxide occupied a volume of 12.0 dm³ at 10.0 °C and a pressure of 105 kPa.

Calculate the amount, in moles, of sulfur dioxide in this sample.

(The gas constant $R = 8.31 \, J \, K^{-1} \, mol^{-1}$)

$$pV = nRT$$

Therefore

$$n = \frac{pV}{RT}$$

SI units must be used:

$p = 105000 \, Pa \quad (105 \times 1000)$

$V = 0.012 \, m^3 \quad (12/1000)$

$T = 283 \, K \quad (10 + 273)$

$$n = \frac{105000 \times 0.012}{8.31 \times 283}$$

$$n = 0.536 \, mol$$

We can also use the ideal gas equation to calculate the volume a gas would occupy at temperatures and pressures other than those at which it was actually measured.

If a gas has a volume $V_1$ measured at pressure $p_1$ and temperature $T_1$, then:

$$\frac{p_1 V_1}{T_1} = nR$$

If the same gas's volume $V_2$ is now measured at a new pressure $p_2$ and temperature $T_2$, then:

$$\frac{p_2 V_2}{T_2} = nR$$

Therefore

$$\frac{p_1 V_1}{T_1} = \frac{p_2 V_2}{T_2}$$

**Worked example**

A sample of gas collected at 30.0 °C and $1.01 \times 10^5$ N m⁻² had a volume of 40.0 cm³. What is the volume of the gas at standard temperature and pressure (stp)?

(At stp, pressure = $1.01 \times 10^5$ N m⁻² and temperature = 273 K)

$$\frac{p_1 V_1}{T_1} = \frac{p_2 V_2}{T_2}$$

$p_1 = 1.01 \times 10^5 \, N \, m^{-2}$      $P_2 = 1.01 \times 10^5 \, N \, m^{-2}$

$V_1 = 40.0 \, cm^3$      $V_2 = ?$

$T_1 = 30 °C = 303 \, K$      $T_2 = 273 \, K$    (Temperature must be in kelvins)

$$\frac{1.01 \times 10^5 \times 40}{303} = \frac{1.01 \times 10^5 \times V_2}{273}$$

$$V_2 = \frac{1.01 \times 10^5 \times 40 \times 273}{1.01 \times 10^5 \times 303}$$

$$V_2 = 36.0 \, cm^3$$

The ideal gas equation is derived from the gas laws.

There are three important ideas describing the behaviour of gases: Boyle's law, Charles' law and Avogadro's principle.

## Boyle's law

In 1662, Robert Boyle published his work on the compressibility of gases. He stated that:

At a constant temperature, the volume of a fixed mass of gas is inversely proportional to its pressure.

It can be written as $V \propto 1/P$ or $PV = $ constant.

Graphs illustrating Boyle's law are shown below:

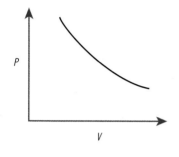 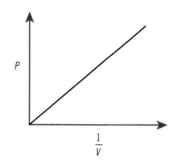

## Charles' law

In 1787, Jacques Charles stated his law on the effect of temperature on the volume of a gas. He stated that:

The volume of a fixed mass of a given gas, at constant pressure, is directly proportional to its temperature in kelvins.

It can be written as $V \propto T$ or $\dfrac{V}{T} = $ constant

A graph illustrating Charles' law is shown below:

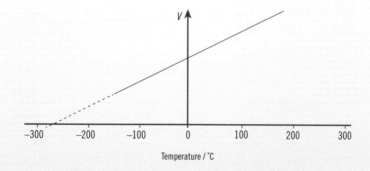

## Avogadro's principle

In 1811, Amadeo Avogadro proposed the following hypothesis about gases:

Equal volumes of different gases, measured at the same temperature and pressure, contain the same number of molecules.

Another way of putting this is that the volume of a gas depends on the amount of moles or $V \propto n$

◀Stretch & challenge

The kinetic theory of gases puts forward a model that explains the gas laws. It makes the general assumption that for an ideal gas:

- The molecules in a gas are in continuous random motion.
- There are no intermolecular forces, so the only interactions between molecules are collisions.
- All collisions are perfectly elastic, i.e. the molecules bounce off each other without their total kinetic energy changing.
- The molecules themselves have no size, i.e. they occupy zero volume.

Can you explain why real gases depart from ideal behaviour at high pressures and low temperatures?

# Concentrations of solutions

The concentration of a solution measures how much of a dissolved substance is present per unit volume of a solution. A solution with a large quantity of solute in a small quantity of solvent is described as concentrated. A solution with a small quantity of solute in a large quantity of solvent is described as dilute.

The concentration of a solution can be stated in many ways, e.g. the concentration of ions in a bottle of mineral water is stated in mg/litre, but the most convenient way is to state the amount in moles of a solid present in 1 dm³ of solution.

i.e. concentration $= \dfrac{\text{amount of moles of solute}}{\text{volume of solution}}$ or $c = \dfrac{n}{V}$

and the unit is moles per cubic decimetre or mol dm⁻³.

This expression can be rearranged in two ways to find the amount in moles or the volume, e.g. 2.1 g of sodium hydrogencarbonate, $NaHCO_3$, is dissolved in 250 cm³ of water. What is the concentration of the solution in mol dm⁻³?

$$\text{Molar mass of } NaHCO_3 = 84$$
$$\text{Amount of moles } NaHCO_3 = \frac{2.1}{84} = 0.025$$
$$\text{Volume of water is } 250\,\text{cm}^3 = \frac{250}{1000} = 0.25\,\text{dm}^3$$
$$\text{Concentration} = \frac{0.025}{0.25} = 0.100\,\text{mol dm}^{-3}$$

Another way of expressing concentration is by using solubility. Solubility is usually measured in g/100g water and since the density of water is 1 g cm⁻³, this is the same as g/100cm³. Therefore to change solubility into concentration in mol dm⁻³ the following steps are used.

**Step 1** Change mass into moles (divide by molar mass).

**Step 2** Change 100 cm³ into dm³ (divide by 1000).

**Step 3** Divide the moles by the volume.

For example, the solubility of potassium nitrate, $KNO_3$, is 31.6 g/100g water. What is the concentration in mol dm⁻³?

**Step 1**     Molar mass $KNO_3 = 101.1$

               Amount in moles $= \dfrac{31.6}{101.1} = 0.313$

**Step 2**     100 cm³ water $= 0.100\,\text{dm}^3$

**Step 3**     Concentration $= \dfrac{0.313}{0.100} = 3.13\,\text{mol dm}^{-3}$

# Acid–base titration calculation

Volumetric analysis is a means of finding the concentration of a solution. An acid–base titration is a type of volumetric analysis. In an acid–base titration, the reacting volumes of an acidic solution and a basic solution are accurately determined. If the concentration of one of these solutions is known, by using the stoichiometric ratios of the solutions, the concentration of the unknown solution can be determined.

Again, there are three main steps to follow:

**Step 1** Find the amount of moles of the solution for which you know the concentration (e.g. acid).

**Step 2** Use a balanced chemical equation to give the stoichiometric (mole) ratio between the acid and base.

**Step 3** Calculate the concentration of the second solution (e.g. base) from the known volume and amount in moles.

## Worked example

A 25.0 cm³ sample of aqueous sodium hydroxide was exactly neutralised by 21.0 cm³ of 0.150 mol dm⁻³ sulfuric acid.

Calculate the concentration of the sodium hydroxide solution in (a) mol dm⁻³ (b) g dm⁻³.

$$H_2SO_4 + 2NaOH \longrightarrow Na_2SO_4 + 2H_2O$$

**Step 1** Amount in moles of $H_2SO_4$ = 0.150 × 0.021 = 3.15 × 10⁻³ mol

(Divide 21.0 cm³ by 1000 to change it into dm³)

**Step 2** From the equation, 1 mol $H_2SO_4$ requires 2 mol NaOH

3.15 × 10⁻³ mol $H_2SO_4$ require 6.30 × 10⁻³ mol NaOH

**Step 3** (a) Concentration of NaOH = $\dfrac{6.30 \times 10^{-3}}{0.025}$ = 0.252 mol dm⁻³

(Divide 25.0 cm³ by 1000 to change it into dm³)

(b) $M_r$ NaOH = 40.0

Concentration NaOH = 40.0 × 0.252 = 10.1 g dm⁻³

## Back-titration

In a back-titration, the amount of excess reactant (e.g. acid) unused at the end of a reaction is found and so the amount used can be calculated. Using the stoichiometric ratios of the acid and base allows us to calculate the exact amount of the second reactant (e.g. base) that has reacted.

## Worked example

A sample containing ammonium sulfate was warmed with 100 cm³ of 1.00 mol dm⁻³ sodium hydroxide solution. After all the ammonia had been evolved, the excess of sodium hydroxide solution was neutralised by 50.0 cm³ of 0.850 mol dm⁻³ hydrochloric acid.

What mass of ammonium sulfate did the sample contain?

The two reactions taking place are

(i) The reaction between the ammonium sulfate and sodium hydroxide

$$(NH_4)_2SO_4(s) + 2NaOH(aq) \longrightarrow 2NH_3(g) + Na_2SO_4(aq) + 2H_2O(l)$$

(ii) The neutralisation of the sodium hydroxide

$$NaOH(aq) + HCl(aq) \longrightarrow NaCl(aq) + H_2O(l)$$

**Step 1** Calculate the amount in moles of HCl used in the neutralisation, this will give the amount in moles of NaOH unused in reaction (i)

Amount in moles of HCl = 0.850 × $\dfrac{50.0}{1000}$ = 0.0425 mol

From equation (ii) mole ratio of HCl : NaOH is 1 : 1

Therefore amount of moles NaOH unused is 0.0425

**Step 2** Calculate the amount in moles of NaOH that reacted with the $(NH_4)_2SO_4$

Initial amount of NaOH = 1.00 × $\dfrac{100}{1000}$ = 0.100 mol

Amount in moles of NaOH used in reaction (i)
= 0.100 − 0.0425 = 0.0575 mol

**Step 3** Calculate the amount in moles of $(NH_4)_2SO_4$ used

From equation (i) mole ratio of NaOH : $(NH_4)_2SO_4$ is 2 : 1

Therefore 0.0575 moles NaOH react with 0.0288 moles $(NH_4)_2SO_4$

**Step 4** Calculate the mass of $(NH_4)_2SO_4$ in the sample

Molar mass of $(NH_4)_2SO_4$ is 132 g mol⁻¹

Mass of $(NH_4)_2SO_4$ = 0.0288 × 132 = 3.80 g

**Study point**

In acid–base titrations there are five things you need to know:

1 The volume of the acid solution.
2 The concentration of the acid solution.
3 The volume of the base solution.
4 The concentration of the base solution.
5 The equation for the reaction.

If you know four of these, you can calculate the fifth.

**Knowledge check** 14

25.0 cm³ of hydrochloric acid are neutralised by 18.5 cm³ of a 0.200 mol dm⁻³ solution of sodium hydroxide. Calculate the concentration of the acid.

**Knowledge check** 15

22.60 cm³ of a 0.0956 mol dm⁻³ solution of nitric acid react with 24.00 cm³ of barium hydroxide solution to form barium nitrate and water. Calculate the concentration of the barium hydroxide.

# Atom economy and percentage yield

When a reaction occurs, the compounds formed, other than the product needed, are a waste. An indication of the efficiency of a reaction can be given as its **atom economy** or its **percentage yield**.

Atom economy is obtained from the chemical equation for the reaction. The higher the atom economy, the more efficient the process.

## Key terms

Atom economy =

$$\frac{\text{mass of required product}}{\text{total mass of reactants}} \times 100\%$$

or

$$\frac{\text{total } M_r \text{ of required product}}{\text{total } M_r \text{ of reactants}} \times 100\%$$

Percentage yield =

$$\frac{\text{mass (or moles) of product obtained}}{\text{maximum theoretic mass (or moles)}}$$

$$\times 100\%$$

## Maths tip

If you're using $M_r$ to calculate atom economy always give your answer to 3 significant figures since all $A_r$ values in the periodic table are given to 3 significant figures.

### Worked example

Titanium used to be manufactured from titanium(IV) oxide by converting titanium(IV) oxide to titanium(IV) chloride then reducing it by magnesium. The process can be represented by:

$$TiO_2 + 2Cl_2 + 2Mg + 2C \longrightarrow Ti + 2MgCl_2 + 2CO$$

Calculate the atom economy for this reaction.

$$\text{Atom economy} = \frac{\text{total } M_r \text{ of required product}}{\text{total } M_r \text{ of reactants}} \times 100$$

$$= \frac{47.9}{79.9 + 142 + 48.6 + 24} \times 100$$

$$= 16.3\%$$

Percentage yield is calculated from the mass of product actually obtained in an experiment.

### Worked example

32.1 g of ethanoic acid are obtained from the oxidation of 27.6 g of ethanol. The reaction can be represented by the equation:

$$C_2H_5OH + 2[O] \longrightarrow CH_3CO_2H + H_2O$$

Calculate the percentage yield for this reaction.

$$\% \text{ yield} = \frac{\text{mass of product obtained}}{\text{maximum theoretical yield}} \times 100$$

To calculate the maximum mass the three steps on page 37 must be followed:

**Step 1**  Calculate the amount, in mol, of 27.6 g $C_2H_5OH$

$$n = \frac{m}{M} = \frac{27.6}{46.0} = 0.600$$

**Step 2**  Use the equation to calculate the amount in moles of $CH_3CO_2H$ formed

1 mol $C_2H_5OH$ gives 1 mol $CH_3CO_2H$

0.600 mol $C_2H_5OH$ give 0.600 mol $CH_3CO_2H$

**Step 3**  Calculate the mass of $CH_3CO_2H$

$$m = nM = 0.600 \times 60.0 = 36.0 \text{ g}$$

$$\% \text{ yield} = \frac{32.1}{36.0} \times 100 = 89.2\%$$

## 16 Knowledge check

Calculate the atom economy for the production of iron in the reduction of iron(III) oxide by carbon monoxide.

$Fe_2O_3(s) + 3CO(g)$
$\longrightarrow 2Fe(s) + 3CO_2(g)$

# Errors or uncertainties

Chemistry and all science depend on quantitative measurements in experiments. The results are expressed in three parts – a number, the units in which the number is given and an estimate of the error in the number.

Basically there are two ways of estimating the error.

If several measurements of a quantity can be made, the values are averaged to obtain a mean value and a standard deviation is calculated. This will not be required in AS Chemistry.

In AS Chemistry the error is estimated from the uncertainties in the apparatus used, such as burette, pipette, balance and thermometer. The size of the uncertainty is determined by the precision of the apparatus.

Uncertainties are usually written with a ± sign. The actual value of the measurement lies somewhere between your reading plus the uncertainty value and your reading minus the uncertainty value.

The uncertainty (error) will be different for different pieces of apparatus. For any piece of apparatus, the uncertainty is taken as one-half of the smallest division on the apparatus, e.g. 0.1 °C on a 0.2 °C thermometer, 0.05 $cm^3$ on a burette (since the scale has marks of 0.1 $cm^3$) and 0.5 mg (0.0005 g) on a three-place balance (since it measures to 0.001 g).

Remember: if you use two readings to work out a measurement, you'll need to combine their uncertainties.

## Percentage error

If you know the error in a reading that you've taken, you can use it to calculate the percentage error in your measurement.

$$\text{percentage error} = \frac{\text{error}}{\text{reading}} \times 100$$

### Worked examples

1. In a titration, the initial reading on a burette is 0.10 $cm^3$ and the final reading is 24.75 $cm^3$. Calculate the percentage error in the titre value.

   The titre value is 24.75 − 0.10 = 24.65 $cm^3$.

   The error on each burette reading is 0.05 $cm^3$. Two readings have been used to find the titre value, so the total error is 0.05 + 0.05 = 0.10 $cm^3$.

   $$\text{percentage error} = \frac{0.10}{24.65} \times 100 = 0.406\%$$

2. In an enthalpy of neutralisation reaction, the temperature of 100 g of solution before the reaction was 18.6 °C and after the reaction was 26.2 °C.

   Calculate the percentage error caused by the thermometer and the error in the evaluation of the heat produced.

   (The thermometer was accurate to ±0.1 °C and the specific heat capacity of the solution was 4.2 $J g^{-1} °C^{-1}$)

   $$\text{Temperature rise} = 26.2 - 18.6 = 7.6 °C$$

   $$\% \text{ error} = \frac{0.20}{7.6} = 2.63\%$$

   $$\text{Heat produced} = 100 \times 4.2 \times 7.6 = 3192 J$$

   Error in value of heat produced = ±2.63% of 3192 J = ±83.9 J

### ≪ Practical check

Mistakes and errors are different.

An error is something that a skilled worker would find difficult to avoid, and is a consequence of the way the apparatus has been constructed and how readings can be made using it.

A mistake is something that a skilled operator can avoid by being careful. Examples of avoidable mistakes in a titration are leaving the funnel in the burette or overshooting the end-point.

### Knowledge check

A student adds a known mass of magnesium to an acid. The balance used to weigh the magnesium can be read to three decimal places. These are his results:

Mass of magnesium and container before adding to acid = 12.362 g

Mass of container after adding magnesium to acid = 12.254 g

Calculate the maximum percentage error in the weighing of the magnesium.

In a volumetric titration we may have percentage errors from the burette, pipette and balance contributing to the total error. Usually at AS it is simpler and satisfactory to identify the largest error and use that on its own. For example, it is very rare for the percentage error in the balance to be significant.

The result of a calculation using several pieces of data cannot have more significant figures than the least number given in any of the terms in the calculation.

For example, solutions A and B react in a $1:1$ mole ratio and $10.00\,cm^3$ of solution B required $25.16\,cm^3$ of solution A, the concentration of which was $0.10\,mol\,dm^{-3}$.

The concentration of B is thus $\dfrac{25.16 \times 0.10}{10.00} = 0.2516\,mol\,dm^{-3}$.

However, since the concentration of A is only known to 2 significant figures, that of B may only be written as $0.25\,mol\,dm^{-3}$.

### Significant figures and decimal places

In calculating the percentage error for the burette in the worked example the calculator gives 0.405679513.

What does this mean? The last five numbers mean nothing; since the error is about 0.4%, that is 4 in a thousand, anything after the third or fourth significant figure has no physical significance. Since the fourth significant figure is 5 or above, the third figure may be rounded up to 0.406.

It is important that full calculator outputs are **not** recorded as FINAL answers, only numbers having physical significance should be used. However, the full output may be used at intermediate stages in a calculation.

A serious error is to destroy information gained in the experiment by over-shortening the result as when a concentration of $0.0946\,mol\,dm^{-3}$ is returned as 0.09 or even 0.1!

These are the rules for working out significant figures:

Zeros to the left of first non-zero digit are not significant, e.g. 0.0003 has one sig. fig.

Zeros between digits are significant, e.g. 3007 has four sig. figs

Zeros to the right of a decimal point with a number in front are significant,

e.g. 3.0050 has five sig. figs.

Decimal places are the number of digits to the right of the decimal point so that 0.044 has 3 decimal places but only two significant figures. Stating the value to two decimal places would give a different value (0.04) than the correct significant figures (0.044).

Always use significant figures and not decimal places when considering errors.

### Standard form and ordinary form

Both forms are commonly used, the standard form being useful when large and small numbers are encountered. It is important that the same number of significant figures is used in both, e.g. ordinary form $0.0052\,mol\,dm^{-3}$, standard form $5.2 \times 10^{-3}\,mol\,dm^{-3}$, both have two significant figures.

# Test yourself

1. Calculate the mass of gold that contains the same number of atoms as there are molecules in 5.75 g of nitrogen dioxide, $NO_2$. [1]

2. Silicon has three stable isotopes $^{28}Si$, $^{29}Si$ and $^{30}Si$.

   (a) The mass spectrum of silicon gave the following results:

   | Isotope | % abundance |
   |---------|-------------|
   | $^{28}Si$ | 92.21 |
   | $^{29}Si$ | 4.70 |
   | $^{30}Si$ | 3.09 |

   (i) State **two** changes which must be made to a sample of solid silicon before it can be investigated in a mass spectrometer. [2]

   (ii) Calculate the relative atomic mass of silicon, giving your answer to **four** significant figures. [2]

   (iii) Explain what is meant by *relative atomic mass*. [1]

   (b) A sample of water containing $^1H$, $^2H$ and $^{16}O$ was analysed in a mass spectrometer. The trace showed a number of peaks. Suggest which ions are responsible for the peaks at mass numbers 1, 18 and 20. [2]

3. An oxychloride of phosphorus which is used to dehydrate amides has a percentage composition, by mass, of P 20.0%; O 10.4%; Cl 69.6%.

   (a) Calculate the **empirical** formula of this compound. [2]

   (b) What other information would you need to determine the **molecular** formula of this compound? [1]

4. Sodium nitrate is widely used in the production of fertilisers.

   (a) Sodium nitrate can be formed by the reaction between sodium carbonate and nitric acid as shown by the equation:

   $Na_2CO_3(aq) + 2HNO_3(aq) \longrightarrow 2NaNO_3(aq) + H_2O(l) + CO_2(g)$

   In an experiment, 25.0 $cm^3$ of a solution of sodium carbonate of concentration 0.0450 mol $dm^{-3}$ required 23.6 $cm^3$ of the acid for complete neutralisation.

   Calculate the concentration of the acid. [3]

   (b) Sodium nitrate decomposes on heating as shown by the equation:

   $2NaNO_3(s) \longrightarrow 2NaNO_2(s) + O_2(g)$

   When a sample was decomposed completely, the oxygen occupied a volume of 700 $cm^3$ at a pressure of 101 kPa and a temperature of 28 °C.

   Calculate the amount, in moles, of oxygen produced. [3]

5. Elinor investigates the reaction between calcium carbonate and dilute hydrochloric acid. She is given 20.0 $cm^3$ of acid of concentration 1.20 mol $dm^{-3}$.

   $CaCO_3(s) + 2HCl(aq) \longrightarrow CaCl_2(aq) + CO_2(g) + H_2O(l)$

   (a) Calculate the minimum mass of calcium carbonate needed to react completely with this amount of acid. [3]

   (b) Calculate the volume of carbon dioxide that would be produced at a temperature of 25 ºC and a pressure of 1 atm. [2]

   (c) Use your answer to (b) to calculate the volume of carbon dioxide that would be produced at 50 °C and 1 atm. [2]

6. Titanium can be manufactured by the Kroll process as shown by the following equations:

   $TiO_2 + 2Cl_2 + C \longrightarrow TiCl_4 + CO_2$ (1)

   $TiCl_4 + 2Mg \longrightarrow Ti + 2MgCl_2$ (2)

   (a) Calculate the atom economy for the production of titanium in equation (2). [2]

   (b) Calculate the maximum mass of titanium which could be obtained from 447 kg of titanium(IV) oxide. [2]

7. Glauber's salt is a form of hydrated sodium sulfate. It can be represented by the formula $Na_2SO_4.xH_2O$.

   When 5.42 g of this hydrate was heated, 2.39 g of the anhydrous salt, $Na_2SO_4$, remained.

   Calculate the value of $x$ in $Na_2SO_4.xH_2O$ [3]

8. Edmund was asked to find the identity of a group 1 metal carbonate. He fully dissolved 0.723 g of the carbonate in water to form a solution. The solution required the addition of 24.80 cm$^3$ of a 0.550 mol dm$^{-3}$ solution of hydrochloric acid for complete neutralisation.

   The equation for the reaction between the metal carbonate and hydrochloric acid is given below. M represents the symbol of the metal:

   $M_2CO_3 + 2HCl \longrightarrow 2MCl + H_2O + CO_2$

   Calculate the relative formula mass of $M_2CO_3$ and hence deduce the metal in the carbonate. [4]

9. When 1 kg of sulfur dioxide is reacted with excess oxygen, 1.225 kg of sulfur trioxide is formed:

   $2SO_2(g) + O_2(g) \longrightarrow 2SO_3(g)$

   Calculate the percentage yield. [3]

10. A 0.920 g sample of impure calcium carbonate was added to 40.0 cm$^3$ of hydrochloric acid of concentration 0.650 mol dm$^{-3}$. At the end of the reaction some acid was left unreacted. The unreacted acid required 25.20 cm$^3$ of sodium hydroxide of concentration 0.325 mol dm$^{-3}$ for complete neutralisation.

    (a) Calculate the number of moles of hydrochloric acid used up in the reaction with impure calcium carbonate. [3]

    (b) Calculate the percentage, by mass, of calcium carbonate in the impure sample. [3]

       The equations for the reactions are

       $CaCO_3 + 2HCl \longrightarrow CaCl_2 + H_2O + CO_2$

       $HCl + NaOH \longrightarrow NaCl + H_2O$

# Bonding

The usefulness of materials depends on their properties, which in turn depend on their internal structure and bonding. By understanding the relationship between these, chemists can design new useful materials. The types of forces between particles are studied and the importance of electrical attractive and repulsive forces stressed.

You should be able to demonstrate and apply knowledge and understanding of the following:

- Ionic bonding in terms of ion formation and the electrical attraction between positive and negative ions.
- Covalent bonding, including coordinate bonding, in terms of the sharing of electrons and the balance of forces of attraction and repulsion within the molecule.
- Ionic and covalent bonds are extremes and that most bonds are intermediate in character.
- Concepts of electronegativity and bond polarity.
- Bonding between molecules is much weaker than the bonds within them.
- Permanent and temporary dipoles and their relative effects on physical properties such as boiling temperatures and solubility.
- Hydrogen bonding and its effect on physical properties such as boiling temperature and solubility.

## Topic contents

### Maths skills >>

- Use angles and shapes in regular 2D and 3D structures.
- Visualise and represent 2D and 3D forms including two-dimensional representations of 3D objects.

# Bonding

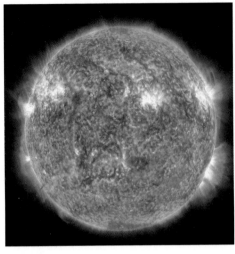

▲ The Sun: 5000°C atoms          ▲ The Earth: 20°C molecules

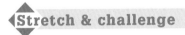
Apart from the inert gases, elements are rarely found as atoms on Earth since they bond together exothermically to form molecules, and temperatures of above a thousand degrees – as in the Sun – may be necessary to break these bonds. In some cases molecules are formed of a few atoms, as in methane, $CH_4$. In other cases millions of atoms may bond together to form a crystal of salt or diamond or metal. Such structures are called giant molecules.

## Chemical bonding

There are different types of **bond: ionic, covalent** (including **coordinate**) and metallic but the basic cause of bonding is the same in each case. The positively charged nuclei and negatively charged electrons are arranged in such a way that the electrostatic attractions outweigh the repulsions.

Bonds may be shown by 'dot and cross' diagrams in which the outer electrons of one atom forming the bond are shown as dots or open circles, and those of the other atom by crosses. The electrons can be drawn in circular orbits.

## Ionic bonding

One atom gives one or more electrons to the other atom and the resulting cation (+) and anion (−) attract one another electrically e.g.

sodium chloride (showing outer electrons only):

$$Na \quad \times Cl \times \longrightarrow [Na]^+ \; [\times Cl \times]^-$$

magnesium fluoride (showing outer electrons only):

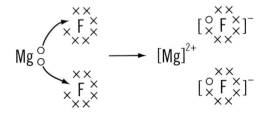

# Covalent bonding

Each atom gives one electron to form a bond pair in which the electron spins are opposed.

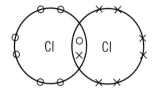

For example, methane, $CH_4$

For example, chlorine, $Cl_2$

# Coordinate bonding

The same as a covalent bond except that both electrons forming the bond pair come from the same atom.

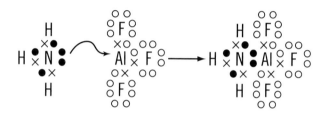

**Stretch & challenge**

It is convenient to show the electrons as dots and crosses to describe these types of bonding but remember that they are really described by orbitals that are volumes enclosing regions of electron density. Bonds are formed by the overlap of these atomic orbitals.

# Attractive and repulsive forces

All bonding results from electrical attractions and repulsions between the protons (in the nucleus) and the electrons, with attractions outweighing repulsions.

In covalent bonds the electrons in the pair between the atoms repel one another but this is overcome by their attractions to BOTH nuclei. If atoms get too close together the nuclei and their inner electrons will repel those of the other atom, so the bond has a certain length. Also, the electron spins must be opposite for the bonds to form.

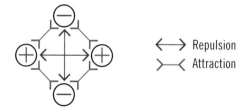

$\longleftrightarrow$ Repulsion

$\rightarrowtail\!\!\!\prec$ Attraction

In ionic bonding, cations and anions are arranged so that each cation is surrounded by several anions and vice versa to maximise attraction and minimise repulsion. Again, repulsions from inner electrons and nuclei prevent the ions from getting too close together.

**Knowledge check**

A bond formed between two atoms in which one atom donates an electron completely to the other atom so that both atoms become charged is called an ............ bond. A bond in which two atoms each donate an electron to form a bond pair is called a ............ bond. A bond in which one atom donates both of the electrons forming the pair bond is called a ............ bond.

**Link**

Ionic solids page 60

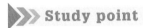

# Electronegativity

Very few compounds are purely ionic or purely covalent. In reality, most compounds show both ionic and covalent properties. The degree of ionic or covalent character depends on the difference in **electronegativity** between the bonded atoms. Electronegativity is measured on the Pauling scale. Some values are given below:

| Li | H | C | N | O | F |
|----|----|----|----|----|----|
| 1.0 | 2.1 | 2.5 | 3.0 | 3.5 | 4.0 |

| Na | Mg | | | S | Cl |
|----|----|----|----|----|----|
| 0.9 | 1.2 | | | 2.5 | 3.0 |

| Cs | | | | | I |
|----|----|----|----|----|----|
| 0.7 | | | | | 2.5 |

This means that fluorine is the most electronegative element. The higher the electronegativity value, the better the element can attract bonding electrons.

## Polar bonds

In a covalent bond, the electron pair is not usually shared exactly evenly between the two atoms. Normally, the atoms have different electronegativities and the bonding electrons are pulled towards the more electronegative atom. This atom will take up a slightly negative charge and the other becomes slightly positive. The partial charges are written above the atoms using the symbol δ (delta).

e.g.

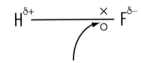

Shared electrons closer to the fluorine since fluorine
is more electronegative than hydrogen

The **bond** is said to be **polar**.

If the two atoms are the same, the atoms have equal electronegativities, so the electrons are equally shared, and the bond is said to be non-polar.

e.g.

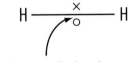

Electrons equally shared

In general, bonds where the electronegativity difference between the two atoms is less than about 0.4 will be non-polar covalent bonds. If the electronegativity difference is between 0.4 and about 1.9 then they will be polar covalent bonds. If the electronegativity difference is about 2.0 or more they will be ionic.

The larger the electronegativity value, the more electronegative the atom. More electronegative atoms are more able to attract electrons within a covalent bond, For example, in methane ($CH_4$) the atoms have relatively similar values so the C–H bond is not highly polar but in hydrogen fluoride (HF) the F atom is more able to attract electrons and therefore the bond is polar.

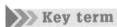

# Forces between molecules

Forces between molecules or **intermolecular** forces are much weaker than bonding within molecules or **intramolecular** bonding. You need to know about three types of intermolecular forces: dipole–dipole forces, induced dipole–induced dipole forces and hydrogen bonds.

## Dipole–dipole forces

Polar molecules have dipoles, one end has a slightly positive charge, the other a slightly negative charge, due to a difference in electronegativities between the atoms in the molecule. If these dipoles arrange themselves so that the negative region of one molecule is close to the positive region of another molecule, there will be an attraction between them. e.g.

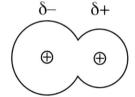

permanent dipole – dipole forces between the HCl molecules

The problem is that the dipoles are not always aligned to produce an attraction between them due to the random movements of the molecules.

## Induced dipole–induced dipole forces

Even molecules with no dipoles show intermolecular bonding. Electrons are in constant motion around the nuclei, therefore at any particular moment the distribution of the electron cloud around the nuclei will not be symmetrical. At any particular moment, the molecule will have a temporary dipole.

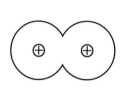

Symmetrical distribution of electron cloud – no dipole

More electrons on the left side – temporary dipole

The δ+ end of the molecule can pull the electron cloud of a neighbouring molecule towards it, giving the left side of that molecule a δ– charge. i.e. it induces a temporary dipole in the neighbouring molecule. The two dipoles are attracted to each other.

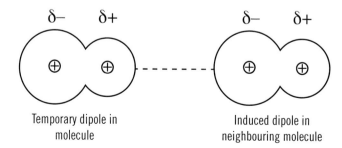

Temporary dipole in molecule

Induced dipole in neighbouring molecule

The second dipole induces a dipole in a third molecule and so on. Since it was the first dipole that led to the induction of the next dipole, the dipoles will always be aligned correctly to produce an attraction between the molecules. The temporary dipoles are being created and destroyed all the time because the electrons are in constant motion. However, this process is very rapid compared with the speed at which the molecules are moving, so as the molecules move, they will continue to be attracted to each other.

**Link**

Volatility of halogens page 70

**Intermolecular bonding** is the weak bonding holding the molecules together, e.g. as in liquids, and governs the physical properties of the substance.

**Intramolecular bonding** is the strong bonding between the atoms in the molecule and governs its chemistry.

The strength of induced dipole–induced dipole forces increases with increasing number of electrons in the molecule. The more electrons in a molecule the greater the fluctuation in the electron cloud around the nuclei and the larger the temporary and induced dipoles created. This means stronger forces between the molecules. This is shown in the trend in boiling temperatures of the noble gases or the halogens.

These two types of intermolecular forces together are called **van der Waals** forces.

# Hydrogen bonds

These are special intermolecular forces that occur between molecules containing hydrogen atoms bonded to small, very electronegative elements which have lone pairs – fluorine, oxygen or nitrogen.

Because fluorine, oxygen and nitrogen are highly electronegative and the hydrogen atom is so small, the $\delta+$ charge in the bonded hydrogen atom is spread over a small volume and so it has a high charge density. The highly polarising $\delta+$ hydrogen atom then attracts a lone pair of electrons from a small highly electronegative atom in another molecule. The $\delta+$ hydrogen atom is sandwiched between two electronegative atoms. It is covalently bonded to one and hydrogen-bonded to the other.

e.g. hydrogen bonding in water

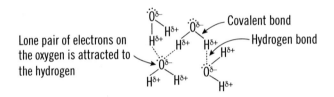

# Effects of hydrogen bonding

### Boiling temperatures

When a liquid boils, energy is needed to overcome all the forces of attraction between the molecules of the liquid. Since hydrogen bonds are stronger than van der Waals forces, molecules that form hydrogen bonds have a higher boiling temperature than molecules of a similar size that cannot hydrogen bond.

Look at the trend in boiling temperatures for the hydrides of the elements of Groups 4, 5 and 6 below.

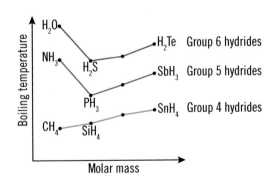

For the Group 4 hydrides, the boiling temperatures increase as you go down the group because the molecules are getting larger with more electrons, and so the van der Waals forces become stronger.

The values for ammonia, $NH_3$, and water, $H_2O$, are much higher than the other hydrides in their groups because extra energy is needed to break the hydrogen bonds between the molecules. The trends in the boiling temperatures for the hydrides of the other elements in Groups 5 and 6 is the same as for the Group 4 hydrides, for the same reasons.

### Solubility

The most significant intermolecular forces between water molecules are hydrogen bonds. Covalent compounds that can replace these bonds by forming new hydrogen bonds with water will dissolve.

# Shapes of molecules and ions

## Valence Shell Electron Pair Repulsion (VSEPR) Theory

This theory states that:

- The shape of a molecule or ion is governed by the number of pairs of electrons in the outer (valence) shell of the central atom.
- The electron pairs arrange themselves around the central atom as far apart as possible from each other so that the repulsion between them is at a minimum.
- Lone pair–lone pair repulsion > lone pair–bonding pair repulsion > bonding pair–bonding pair repulsion.

The first two points are used to find the basic shape of the molecule or ion. The third point is used to estimate values for the bond angles.

### How to find the number of electron pairs

To work out the shape of a molecule or ion you need to know the number of bonding pairs and lone pairs around the central atom.

You can do this by drawing a dot and cross diagram or by following these steps:

1. Write down the number of electrons in the outer shell of the central atom. (This will be the same as the periodic table group number.)
2. Add one electron for each bond being formed. (The formula will give the number of bonds.)
3. Allow for any ion charge. (Add one electron if the ion has a charge of 1–. For a 1+ charge, deduct one electron.)
4. Divide the total number of electrons by 2 to find the number of electron pairs.
5. Compare the number of electron pairs with the number of bonds to find the number of bonding pairs and lone pairs. (Use the formula to find the number of bonding pairs.)

**Study point**

- Electron pairs may be shared (bonding pairs) or unshared (lone pairs).
- Since electrons are negatively charged, all electron pairs repel each other.
- Bonding pairs of electrons are spread out between the two atoms bonded in the molecule but lone pairs stay close to the central atom so repel more.

**Study point**

Remember, for step 2, each double bond donates two electrons and, for step 5, a double bond counts as two bonds.

## Worked examples

Hydrogen sulfide, $H_2S$

1. Sulfur is in group 6 so has 6 electrons in its outer shell.
2. Sulfur forms two bonds with hydrogen so 2 electrons are added from the hydrogen atoms.
3. $H_2S$ is a molecule, not an ion, so no charge.
4. The total number of electrons is 8, therefore 4 electron pairs.
5. 2 electron pairs are involved in bonding with the hydrogen atoms, so there are 2 lone pairs.

Ammonium ion, $NH_4^+$

1. Nitrogen is in group 5, so has 5 electrons in its outer shell.
2. Nitrogen forms four bonds with hydrogen so 4 electrons are added from the hydrogen atoms.
3. The ion has a charge of 1+ because it has lost an electron, so subtract 1 electron.
4. The total number of electrons is 8 giving 4 electron pairs.
5. All the electron pairs are bonding because of the four hydrogens.

## Working out the shape of molecules and ions

Once you know how many bonding pairs and lone pairs of electrons are around the central atom, you can work out the shape.

### Two electron pairs around the central atom

The two bonding pairs of electrons arrange themselves at 180° to each other because that's as far apart as they can get. A molecule is then described as being **linear**,

e.g. carbon dioxide

Double bonds are treated as single units. Carbon dioxide, $CO_2$, has two double bonds so although it contains four bonding pairs of electrons, each double bond uses two bond pairs. The two double bond units arrange themselves as far apart as possible so the molecule is linear.

### Three electron pairs around the central atom

The three bonding pairs of electrons arrange themselves as far apart as possible, so they all lie in the same plane at 120° to each other. This arrangement is called **trigonal planar**,

e.g. boron trifluoride

### Four electron pairs around the central atom

**Four bonding pairs**

The electron pairs all repel each other equally. All the bond angles are 109.5° and the shape of the molecule is **tetrahedral**. A tetrahedron is a regular triangular-based pyramid,

e.g. methane

The carbon atom would be at the centre and the hydrogens at the four corners.

**Exam tip** »

You must be able to draw a 3D representation of the shapes of molecules with four or more electron pairs. Straight lines show bonds that are flat against the page, broken lines show bonds that point into the page and wedges show bonds that stick out of the page.

### Three bonding pairs and one lone pair

Again, the electron pairs arrange themselves in a tetrahedral shape, but there is more repulsion between a lone pair and a bonding pair than between two bonding pairs. This forces the bonding pairs together, slightly reducing the bond angle between them to 107° and the shape of the molecule is **trigonal pyramidal**,

e.g. ammonia

Although the electron pair arrangement is tetrahedral, when you describe the shape you only take notice of the atoms. The ammonia is like a pyramid with the three hydrogens at the base and the nitrogen at the top.

### Two bonding pairs and two lone pairs

This also takes a tetrahedral arrangement, but the lone pair–lone pair repulsion forces the bonding pairs even closer and reduces the angle between them to 104.5°. The shape is described as **bent** or **V-shaped**.

e.g. water

### Exam tip

You do not need to know the bond angles in pyramidal or bent shape molecules, only that they are less than in tetrahedral molecules.

### *Five electron pairs around the central atom*

Some central atoms can expand the octet, i.e. they can have more than eight bonding electrons. In this case there are 10 electrons, consisting of 5 bonding pairs, around the central atom. Repulsion between the bonding pairs means that three of the atoms are in a plane at 120° to each other and the other two atoms are at right angles to this plane. The shape is described as **trigonal bipyramidal**,

e.g. phosphorus pentachloride

### Exam tip

You only need to know the name of this shape. You will not be required to draw it or know the bond angles.

### *Six electron pairs around the central atom*

There are 12 electrons, consisting of 6 bonding pairs, around the central atom. They arrange themselves entirely at 90° in a shape described as **octahedral**,

e.g. sulfur hexafluoride

The angle between the bonds of two fluorine atoms opposite one another is 180°.

The table below gives a summary of the shapes and bond angles.

| Number of bonding pairs | Number of one pairs | Examples | Shape | Bond angles |
|---|---|---|---|---|
| 2 | 0 | $BeCl_2$, $CO_2$ | Linear | 180° |
| 3 | 0 | $BF_3$, $AlCl_3$ | Trigonal planar | 120° |
| 4 | 0 | $CH_4$, $NH_4^+$ | Tetrahedral | 109.5° |
| 3 | 1 | $NH_3$, $H_3O^+$ | Trigonal pyramidal | 107° |
| 2 | 2 | $H_2O$, $F_2O$ | Bent or V-shaped | 104.5° |
| 5 | 0 | $PCl_5$ | Trigonal bipyramidal | 120°, 90° |
| 6 | 0 | $SF_6$ | Octahedral | 90° |

### Knowledge check 3

(a) Give the shapes of the following molecules:
$BeI_2$, $SO_3$, $CCl_4$, $PF_5$
($SO_3$ only contains double bonds)

(b) Match the following molecules with the bond angles listed:
$BeH_2$, $BCl_3$, $CF_4$, $PF_5$, $SF_6$
90°, 109.5°, 120°, 180°

(c) State whether $BCl_3$ and $PCl_3$ have the same shape. Explain your answer.

# Test yourself

1. Using **outer** electrons only, draw a dot and cross diagram to show the formation of the bonding in strontium bromide. [2]

2. The electronegativity values of some elements are given below:

| Atom | H | O | F | S | Cl | Ga | Br |
|---|---|---|---|---|---|---|---|
| Electronegativity value | 2.1 | 3.5 | 4.0 | 2.5 | 3.0 | 1.6 | 2.8 |

   (a) Use this data to identify any dipoles present in the following bonds. Mark their polarity clearly.

   H–Br                    Cl–Cl                    S–O [2]

   (b) Use the data to explain why gallium chloride is considered to be a covalent compound while gallium fluoride is an ionic compound. [2]

3. Oxygen difluoride has the formula $OF_2$.

   (a) Indicate the polarity in the O–F bond by use of the symbols $\delta^+$ and $\delta^-$. Give a reason for your answer. [2]

   (b) Using **outer** electrons only, draw a dot and cross diagram to illustrate the bonding in oxygen difluoride, $OF_2$. [1]

4. Describe the formation and nature of a covalent bond. [3]

5. The bond angles in molecules of methane and water are shown in the diagrams below:

   Using the valence shell electron pair repulsion (VSEPR) theory:

   (a) State why methane has the shape shown. [2]

   (b) Explain why the H–O–H bond angle in water is less than the H–C–H bond angle in methane. [2]

6. Select all the molecules from the list below that have bond angles of less than 109°.

   $BCl_3$  $H_2O$  $NH_4^+$  $PH_3$  $SF_6$ [2]

7. The boiling temperatures of some hydrogen halides are shown in the table below:

| Hydrogen halides | Boiling temperature / K |
|---|---|
| HF | 293 |
| HCl | 188 |
| HBr | 206 |
| HI | 238 |

   Explain, in terms of the nature of the intermolecular bonding present, why the boiling temperature of hydrogen iodide is higher than those of hydrogen chloride and hydrogen bromide, but is lower than that of hydrogen fluoride. [4]

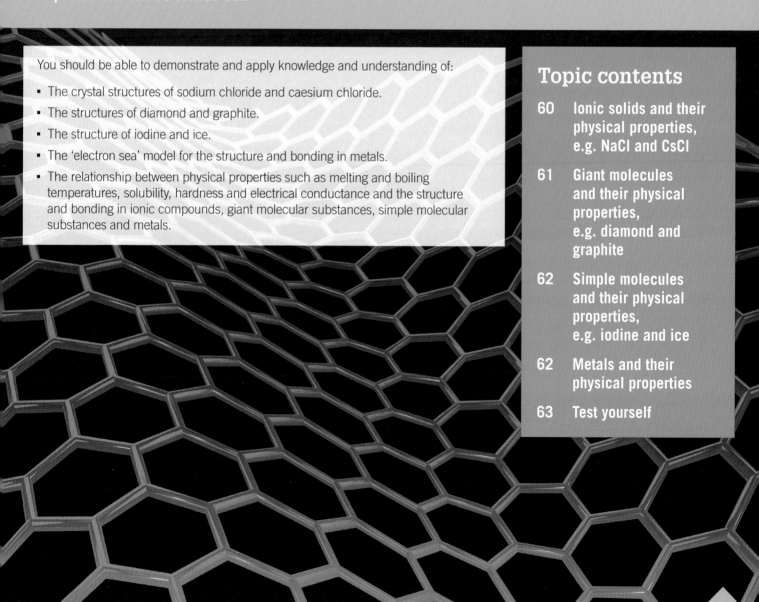

# Solid structures

The properties and usefulness of solid materials depend very much on their structure and bonding at the molecular level. The more that these are understood, the more successful research into design and improvement can be carried out.

You should be able to demonstrate and apply knowledge and understanding of:

- The crystal structures of sodium chloride and caesium chloride.
- The structures of diamond and graphite.
- The structure of iodine and ice.
- The 'electron sea' model for the structure and bonding in metals.
- The relationship between physical properties such as melting and boiling temperatures, solubility, hardness and electrical conductance and the structure and bonding in ionic compounds, giant molecular substances, simple molecular substances and metals.

## Topic contents

# Ionic solids

Ionic solids (crystals) are giant lattices of positive and negative ions. The structure of the crystal depends on the relative number of ions and their sizes. Structures are made of the same base unit repeated over and over again. The ions are arranged in such a way that the electrostatic attraction between the oppositely charged ions is greater than the electrostatic repulsion between ions with the same charge.

We shall consider two basic structures in which the ratio is 1:1.

## Sodium chloride

Sodium chloride consists of a regular array of sodium ions, $Na^+$, and chloride ions, $Cl^-$.

Each $Na^+$ ion is surrounded by six $Cl^-$ ions and vice versa. The **coordination number** of each ion is 6.

**Study point**

In reality, the ions in NaCl are touching one another.

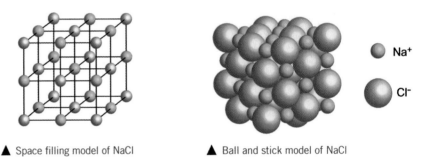

▲ Space filling model of NaCl          ▲ Ball and stick model of NaCl

## Caesium chloride

**Knowledge check**

Why is it misleading to refer to a molecule of caesium chloride?

Caesium chloride is an example of the simplest structure found in ionic compounds. The caesium ion, $Cs^+$, is larger than the $Na^+$ ion therefore more $Cl^-$ ions can fit around it.

Each $Cs^+$ ion is surrounded by eight $Cl^-$ ions and vice versa. The coordination number of each ion is 8.

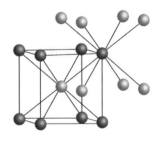

▲ Diagram of CsCl structure

## Physical properties

The physical properties of an ionic solid are decided by the structure and bonding of ionic compounds. Typical characteristics are:

- High melting temperatures. The giant lattices are held together by strong electrostatic forces between the oppositely charged ions. It takes a large amount of energy to overcome these forces of attraction.
- Often soluble in water. Water molecules are polar – the oxygen atoms have a partial negative charge and the hydrogen atoms a partial positive charge. In solution the oxygen ends of the water molecules are attracted to the positive ions and the hydrogen ends of the water molecules are attracted to the negative ions.

If the attraction between water molecules and ions is comparable with the energy required to separate the ions, the water molecules pull the ions away from the lattice and cause the solid to dissolve.

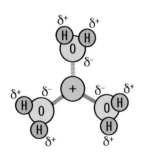

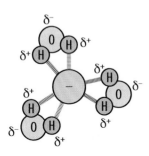

▲ Ionic compounds in water

- Hard but brittle. When force is applied, layers of ions slide over each other causing ions of the same charge to be next to each other. The ions repel each other and the crystal shatters.
- Poor electrical conductivity when solid, but good when molten or dissolved. An electric current will flow if charged particles are free to move when a potential difference is applied. In the solid state the ions are fixed in position by the strong ionic bonds. However, when molten or dissolved, the ions are free to move and will move to the electrode of opposite sign, so will carry the current.

# Giant covalent solids

Giant covalent solids consist of networks of covalently bonded atoms that stretch throughout the whole structure. They are sometimes called macromolecules. Carbon (and silicon) atoms can form this type of structure. Different forms of the same element in the same state are called allotropes. Two allotropes of carbon are diamond and graphite.

## Diamond

In diamond, each carbon atom is covalently bonded to four others. The atoms arrange themselves in a tetrahedral shape. The bonding forces are uniform throughout the structure.

### Physical properties
- Very high melting temperature. The energy needed to break the strong covalent bonds is very high.
- Extremely hard. This is due to the strength of the covalent bonds and the geometrical rigidity of the structure. Many cutting tools are tipped with diamonds.
- Insoluble in water. There are no ions to attract the polar water molecules.
- Poor conductor of electricity. There are no free electrons or ions present.

▲ Structure of diamond

>> **Key term**

**Delocalised** means that an electron is not attached to a particular atom – it can move around between atoms.

## Graphite

Graphite consists of layers of hexagonal rings. In a layer, each carbon is joined to three others by strong covalent bonds. The fourth electron from each carbon atom is **delocalised** within the layer. The hexagonal layers are held together by weak van der Waals forces (induced dipole – induced dipole forces).

Delocalised means that an electron is not attached to a particular atom – it can move around between atoms.

### Physical properties
- Very high melting temperature. It has strong covalent bonds in the hexagon layers.
- It has a soft, slippery feel. The weak forces between the layers are easily broken, so the layers can slide over each other. It is used as a lubricant.
- Insoluble in water. There are no ions to attract the polar water molecules.
- Good conductor of electricity. The delocalised electrons are free to move along the layers so an electric current can flow. The delocalised electrons are not free to move from one layer to the next, therefore it can only conduct parallel to its layers.
- Low density. There is a relatively large amount of space between the layers because the length of the covalent bonds in the layers is much shorter than the length of the van der Waals forces between the layers. It is used to make strong, lightweight sports equipment.

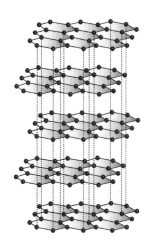

▲ Structure of graphite

**Knowledge check** 2

a) Why can graphite conduct electricity but diamond cannot?

b) Why is graphite used in pencils?

# Simple molecular solids

Simple molecular solids have covalent bonds within molecules held together by weak intermolecular forces.

## Physical properties

- Low melting and boiling temperatures. Although the covalent bonds within the molecules are strong, the intermolecular forces holding the molecules together are weak and do not need much energy to break.
- Soft. The weak intermolecular forces between the molecules are easily broken.
- Normally insoluble in water. There are no ions to attract the polar water molecules. However, compounds that can form hydrogen bonds with water will be soluble.
- Poor conductors of electricity. They do not contain delocalised electrons or ions.

Two common examples of simple molecular solids are iodine and ice.

## Iodine

Atoms are covalently bonded in pairs to form diatomic $I_2$ molecules. These molecules are held together by weak van der Waals forces and are arranged in a regular pattern.

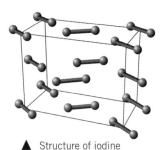

▲ Structure of iodine

## Ice

In ice, molecules of water are arranged in rings of six held together by hydrogen bonds.

In this ordered structure, the water molecules are further apart than they are in the liquid state. The structure creates large areas of open space inside the rings and as a result, at 0°C, ice is less dense than liquid water.

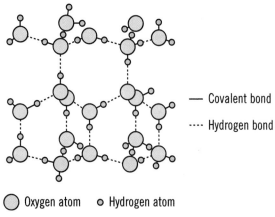

— Covalent bond

···· Hydrogen bond

◯ Oxygen atom   ○ Hydrogen atom

▲ Structure of ice

# Metals

When metal atoms are close to each other, each atom loses control over its outer electrons. These electrons are no longer restricted to a particular metal atom but are delocalised, i.e. free to move throughout the metal. This leaves a positive cation e.g. $Na^+$, $Mg^{2+}$.

Metals, therefore, consist of a regular arrangement of metal cations (a lattice) surrounded by a 'sea' of delocalised electrons. There are electrostatic forces of attraction between the nucleus of the cations and the delocalised electrons. This is known as metallic bonding.

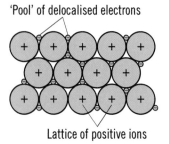

'Pool' of delocalised electrons

Lattice of positive ions

▲ Metal structure

### Knowledge check

**3** 

A compound has a melting temperature of 146°C, is soluble in water and does not conduct electricity when solid or dissolved. State, giving a reason, its type of structure.

### Exam tip

You will not be expected to draw this structure in an exam.

### Link

Hydrogen bonding page 54

# Physical properties

- High melting temperatures. A large energy is needed to overcome the strong forces of attraction between the nuclei of the metal cations and the delocalised electrons. The melting temperature is affected by the number of delocalised electrons per cation and the size of the cation.

- Hard. The metallic bond is very strong.

- Insoluble in water. There are no ions to attract the polar water molecules. (Each metal cation represents the rest of the atom apart from the outer electron. That electron has not been lost, it is still there in the structure.)

- Good conductors of electricity both in the solid and molten state. The delocalised electrons can carry a current because, when a potential difference is applied across the ends of a metal, the electrons will be attracted to, and move towards, the positive terminal of the cell. They are also good thermal conductors because the delocalised electrons can pass kinetic energy to each other.

- Malleable (can be shaped) and ductile (can be drawn into a wire). When a force is applied to a metal the layers of cations can slide over each other. However, the delocalised electrons move with the cations and prevent forces of repulsion forming between the layers.

**Knowledge check** 4

a) Name the type of bonding found in (i) calcium, (ii) calcium chloride.

b) Explain why aluminium can be hammered into different shapes.

# Test yourself

1. Potassium iodide has the same crystal structure as sodium chloride.

   Label the ions present in the diagram of potassium iodide shown below.

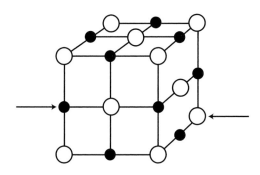

   [1]

2. Both graphite and iodine contain covalent bonding.

   Explain why solid iodine can be converted into a vapour at a much lower temperature than graphite.          [2]

3. An element has a melting temperature of 39 °C and conducts electricity when it is a solid or a liquid.

   State the type of bonding present in the structure and explain why it can conduct electricity.          [3]

4. Although diamond is a non-metal, it can show similar physical properties to metals.

   Describe the structure and bonding in diamond and in metals. State one physical property which is common to both diamond and metals. Relate this property to the structures and bonding you have described.          [5]

# The periodic table

The periodic table is an essential tool in dealing with the vast number of chemical elements and compounds that exist, enabling us to see trends and make predictions. In this topic we see how the table is set up and operates, and study the important chemistry of the s-block Group 1 and 2 elements and compounds and the halogens in Group 7.

## Topic contents

You should be able to demonstrate and apply knowledge and understanding of:

- The arrangement of elements in the table in terms of the electronic structures in the s, p and d blocks.
- Oxidation and reduction in terms of electron transfer and oxidation number.
- The trends in ionisation energy, electronegativity and melting temperatures across periods and down groups.
- The common reactions of the s-block elements with water, acids and oxygen, and trends in their reactivity.
- Characteristic flame colours shown by Group 1 and Group 2 compounds.
- The common reactions of Group 2 ions with hydroxide, carbonate and sulfate ions and the thermal stability and solubility trends of the common salts.
- The trend in volatility of the halogens.
- The trends in reactivity and the relative oxidising powers of the halogens as shown by their reactions with metals and displacement reactions.
- The reaction of halide ions with $Ag^+$ followed by dilute ammonia.
- The use of chlorine and fluoride ions.
- How to perform simple gravimetric analysis and salt formation and crystallisation.

### Maths skills 》

- Translate information between graphical and numerical forms.
- Use an appropriate number of significant figures.
- Change the subject of an equation.
- Substitute numerical values into algebraic equations.

# The structure of the periodic table

The periodic table arranges the elements according to increasing atomic number.

The vertical columns are called **groups**. All the elements in the eight main groups contain the same outer electron configuration. The group number tells you the number of electrons in the outer shell, e.g. Group 1 elements have one electron in their outer shell. Because of this, elements within a group have similar chemical properties.

The horizontal rows are called **periods**. All the elements in a period have the same number of quantum shells containing electrons, e.g. all the elements in period 2 have electrons in both first and second quantum shells.

The table is also divided into **blocks**. Groups 1 and 2 are in the s-block, since the elements' outer electrons are in an s orbital. Groups 3 to 8 are in the p-block, since the elements' outer electrons are in a p orbital. The elements between Groups 2 and 3 are in the d-block, since the elements' outer electrons are in a d orbital.

Usually metals are on the left and middle of the periodic table and non-metals are on the right.

# Trends in physical properties

## Ionisation energy

There is a general increase in ionisation energy across a period. There is an increase in nuclear charge in the same energy level, so there is little extra shielding and therefore a greater attraction between the nucleus and outer electrons.

There is a decrease between Group 2 and Group 3. The Group 3 elements' outer electron is in a new subshell of slightly higher energy level and is partly shielded by the s electrons.

There is a decrease between Group 5 and Group 6. In Group 6 the electron is removed from an orbital containing a pair of electrons. The repulsion between these electrons makes the electron easier to remove. In Group 5, the electron is removed from a singly occupied orbital.

Ionisation energy decreases down a group. The outer electron has increased shielding from inner electrons and is further from the nucleus. This outweighs the increase in nuclear charge.

## Electronegativity

Electronegativity is a measure of the tendency of an atom to attract a bonding pair of electrons.

Electronegativity increases across a period. There is an increase in nuclear charge, but the bonding electrons are always shielded by the same inner electrons, so there is a greater attraction between the nucleus and the bonding pair.

Electronegativity decreases down a group. The bonding electrons have increased shielding from the nucleus, so the attraction between the nucleus and bonding electrons decreases.

Therefore, the most electronegative elements are at the top on the right-hand side of the periodic table, while the least electronegative elements are at the bottom on the left-hand side.

### ⟫ Study point

At the beginning of the 19th century, scientists realised that some elements showed similar properties and began to arrange them in groups called triads. By the mid 19th century, the development of scientific knowledge led to the law of octaves, i.e. every eighth element shows similar properties. In 1869 Mendeleev produced the periodic table. He left gaps in the table and predicted the properties of unknown elements which would fill the gaps. When these elements were discovered, they had the properties predicted by Mendeleev. The regular trends in properties down and across the groups led to a great simplification in the chemistry of the elements and this guided the development of the electronic structure advances at the beginning of the 20th century.

**Link**

See the periodic table on page 225

### ⟫ Study point

Ionisation energy has been fully covered in Unit 1.2 pages 24–25, but here is a recap.

### Knowledge check 1

Draw a sketch to show the trend in ionisation energy for the first 10 elements in the periodic table.

**Link**

1.4 Electronegativity page 52

### Knowledge check 2

Explain why fluorine has a larger electronegativity value than nitrogen.

## Melting and boiling temperatures

To understand the trend in melting and boiling points in the periodic table, you need to understand the structure and bonding of the elements.

Consider the melting and boiling temperatures of the period 3 elements:

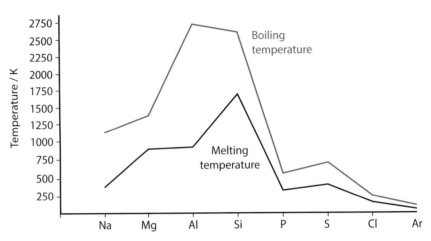

There is a general increase from the first to the fourth element, a large decrease to the fifth element then a small general decrease to the eighth element.

The structure of the elements changes across the period from metallic to giant covalent and then to simple molecular.

Sodium, magnesium and aluminium have metallic bonding. There is an increase because the metallic bonding gets stronger. The metal ions have a greater charge and there is an increased number of delocalised electrons.

Silicon has a giant covalent structure. Each atom is bonded covalently to four other atoms (similar to diamond). A large amount of energy is needed to break all these bonds.

Phosphorus, sulfur and chlorine are simple molecular substances. Although the covalent bonds between the atoms in the molecules are strong, the intermolecular forces holding the molecules together are weak and do not need much energy to break.

Argon has the lowest melting and boiling temperatures because it exists as separate atoms held together by very weak induced dipole–induced dipole forces.

The trends in melting and boiling temperatures are similar in period 2, but boron, in Group 3, has a giant covalent structure not a metallic structure.

1.5 Solid structures pages 61–62

# Reduction and oxidation (Redox)

Reduction and oxidation take place together:

e.g. $Mg + CuO \longrightarrow MgO + Cu$

Magnesium has gained oxygen so is oxidised.

Copper oxide has lost oxygen so is reduced.

To explain the reaction in terms of **electron transfer** we use half equations.

Magnesium loses electrons: $Mg \longrightarrow Mg^{2+} + 2e$

Copper ions gain electrons: $Cu^{2+} + 2e \longrightarrow Cu$

**Knowledge check**

Explain, in terms of electron transfer, why $Ca + 2HCl \longrightarrow CaCl_2 + H_2$ is classified as a redox reaction.

Therefore, oxidation is loss of electrons (Mg is oxidised) and reduction is gain of electrons ($Cu^{2+}$ is reduced). The $Cu^{2+}$ is the oxidising agent (it oxidises the Mg) and is itself reduced in the process (forms Cu). The Mg is the reducing agent (reduces the $Cu^{2+}$) and is itself oxidised in the process (forms $Mg^{2+}$).

Another way to tell if a reaction is a redox reaction is to work out the **oxidation numbers** of the atoms or ions.

If the oxidation number increases, the species is oxidised; if it decreases, the species is reduced:

e.g. $\quad Ba + Cl_2 \longrightarrow BaCl_2$

The oxidation number of barium has increased from 0 to +2, therefore it has been oxidised.

The oxidation number of chlorine has decreased from 0 to –1, therefore it has been reduced.

The oxidation number of an atom does not always change when it reacts. It can be helpful to write the oxidation numbers of each atom underneath the symbol in an equation to see what the changes are, if any:

e.g. $\qquad\qquad 2HNO_3 + 6HI \longrightarrow 2NO + 3I_2 + 4H_2O$

oxidation number $\quad$ +1+5–2 $\quad$ +1–1 $\qquad$ +2–2 $\quad$ 0 $\qquad$ +1–2

The nitrogen has been reduced, oxidation number decreases from +5 to +2.

The iodide has been oxidised, oxidation number increases from –1 to 0.

(The hydrogen and oxygen have not changed.)

**Knowledge check 4**

For the reaction:
$Sr + 2H_2O \longrightarrow Sr(OH)_2 + H_2$

Use oxidation numbers to state which species has been oxidised and which reduced.

**Link**

Oxidation numbers page 13

# Chemistry of Group 1 and Group 2 metals and their compounds (the s-block elements)

## Reaction with water

Group 1 metals react vigorously with cold water to form the hydroxide and hydrogen:

e.g. $\quad 2Na + 2H_2O \longrightarrow 2NaOH + H_2$

The reaction increases in vigour as you go down the group:

- Lithium floats on the water, gently fizzing.
- Sodium melts into a ball that dashes around the surface.
- Potassium melts into a ball and catches fire.
- Caesium explodes and shatters the glass container.

▲ Li + $H_2O$

▲ Cs + $H_2O$

Group 2 metals react less vigorously, in fact magnesium only reacts very slowly. Again, the hydroxide and hydrogen are formed:

e.g. $Ca + 2H_2O \longrightarrow Ca(OH)_2 + H_2$

The reactivity increases as you go down the group:

- Calcium produces a steady stream of bubbles and the liquid goes cloudy as a white precipitate of calcium hydroxide forms.

- Barium produces greater effervescence and the solution is clearer since barium hydroxide is more soluble.

▲ $Ca + H_2O$     ▲ $Ba + H_2O$

Magnesium reacts with steam to produce the oxide and hydrogen.

# Trends in reactivity

Reactivity increases as you go down a group. The reason for this is because when the s-block metals react, they lose electrons to form positive ions. Since ionisation energies decrease down a group, the energy needed to form positive ions decreases. This leads to lower activation energies and therefore faster reactions. Group 1 metals are more reactive than Group 2 metals because Group 1 metals lose only one electron while Group 2 metals lose two electrons.

### Reaction with acids
All the Group 2 metals react vigorously with hydrochloric acid to produce a colourless solution of the metal chloride and bubbles of hydrogen:

e.g. $Sr + 2HCl \longrightarrow SrCl_2 + H_2$

The reactivity increases as you go down the group.

Of Group 2 metals only magnesium reacts with sulfuric acid as the other members have insoluble sulfates.

Group 1 metals are too reactive to be added directly to acids.

### Reaction with oxygen
Apart from magnesium, all Group 2 metals tend to burn with a characteristic flame. All Group 2 metals burn to form solid white oxides:

e.g. $2Mg + O_2 \longrightarrow 2MgO$

The Group 1 metals also form white solids and burn with a characteristic flame:

e.g. $4Li + O_2 \longrightarrow 2Li_2O$

The Group 1 metals also form peroxides and superoxides but you do not need to know about these in this course.

# Oxides and hydroxides

In general, metal oxides are basic and non-metal oxides are acidic.

All the s-block metal oxides are strong bases and they neutralise acids to form a salt and water:

e.g.    $MgO + 2HCl \longrightarrow MgCl_2 + H_2O$

The Group 1 oxides and barium oxide react with water to form a soluble hydroxide:

e.g.    $Na_2O + H_2O \longrightarrow 2NaOH$

Since the hydroxides are soluble, they are alkalis.

The other Group 2 hydroxides are not very soluble, so saturated solutions of these hydroxides are only weakly basic because the concentration of hydroxide ion is very low.

# Test for cations

All s-block elements (apart from Mg) may be identified by a flame test. A clean metal wire (or splint) is moistened with hydrochloric acid, dipped in the compound and held in a non-luminous Bunsen flame.

The characteristic colours for the metal ions are:

| Ion | Colour | Ion | Colour |
|-----|--------|-----|--------|
| $Li^+$ | red | $Mg^{2+}$ | (no colour) |
| $Na^+$ | orange-yellow | $Ca^{2+}$ | brick red |
| $K^+$ | lilac | $Sr^{2+}$ | crimson |
|  |  | $Ba^{2+}$ | apple green |

# Solubility in water

All Group 1 compounds are soluble. However, many Group 2 compounds are not. Here are some trends for Group 2 compounds:

- All **nitrates** are **soluble**.
- All **carbonates** are **insoluble**.
- The **hydroxides** become **more soluble** as you go down the group. Therefore, magnesium hydroxide is insoluble, while barium hydroxide is soluble.

    $Mg^{2+}(aq) + 2OH^-(aq) \longrightarrow Mg(OH)_2(s)$

- The **sulfates** become **less soluble** as you go down the group. Therefore, magnesium sulfate is soluble but barium sulfate is insoluble.

    $Ba^{2+}(aq) + SO_4^{2-}(aq) \longrightarrow BaSO_4(s)$

- All the precipitates are white.
- These trends can be used to distinguish between unknown solutions containing Group 2 cations.

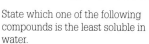

**Knowledge check 5**

State which one of the following compounds is the least soluble in water.

Barium hydroxide, calcium carbonate, magnesium sulfate, sodium carbonate.

### 6 Knowledge check

Describe a chemical test, apart from a flame test, which would distinguish between unlabelled aqueous solutions of barium chloride and magnesium chloride.

## Thermal stability of hydroxides and carbonates

All Group 2 hydroxides decompose on heating to the oxide and steam:

e.g.     $Ca(OH)_2(s) \longrightarrow CaO(s) + H_2O(g)$

The thermal stability increases as you go down the group, i.e. the hydroxides have to be heated more strongly before they will decompose.

All Group 2 carbonates decompose on heating to the oxide and carbon dioxide:

e.g.     $MgCO_3(s) \longrightarrow MgO(s) + CO_2(g)$

Again, the thermal stability increases as you go down the group.

This trend can be shown in the laboratory by heating the carbonate and seeing how long it takes the $CO_2$ formed to turn limewater cloudy.

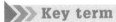

### Study point

You do not study fluorine because it sometimes behaves differently from chlorine, bromine and iodine, or astatine since it only exists as radioactive isotopes.

### Link

Forces between molecules page 53

### Key term

**Volatility** describes how readily a substance vaporises.

# Chemistry of Group 7 halogens and halides

The elements are known as halogens because they all form salts called halides. The term 'halogen' comes from the Greek and means 'salt producer'. The halogens exist as diatomic molecules containing a single covalent bond, e.g. $Cl_2$,

$$:\overset{..}{\underset{..}{Cl}}:\overset{..}{\underset{..}{Cl}}:$$

At room temperature chlorine is a green gas, bromine a red-brown liquid and iodine a grey solid.

As the number of electrons increases with atomic number, there is an increase in the induced dipole–induced dipole intermolecular forces holding the diatomic molecule together. Therefore, the melting and boiling temperatures increase as you go down the group. Substances that form vapours easily are called volatile. A substance with a low boiling temperature has a high **volatility**. Consequently, volatility decreases down the group.

## Trends in reactivity

The halogens react by gaining electrons to form negative halide ions. Since they gain electrons during reactions, halogens are reduced, and they oxidise the other substance. As you go down the group, the outer electrons are shielded more and are further from the nucleus. So, it gets harder to attract electrons, and both reactivity and oxidising power decrease down the group.

The halogens react directly with most metals to form the halide.

For example, sodium burns in a jar of chlorine gas forming white sodium chloride:

$$2Na + Cl_2 \longrightarrow 2NaCl$$

Iron wool also burns directly in chlorine or bromine vapour to give the iron(III) halide. However, when it burns in iodine vapour it only produces iron(II) iodide since iodine is less reactive and is a weaker oxidising agent.

$2Fe + 3Br_2 \longrightarrow 2FeBr_3$     The bromine oxidises the iron to the +3 oxidation state.

$Fe + I_2 \longrightarrow FeI_2$     The iodine oxidises the iron to the +2 oxidation state.

# Displacement reactions

A halogen in a higher position in the group will oxidise a halide ion from lower in the group since oxidising powers decrease down the group. Therefore, when a halogen is added to an aqueous solution containing a halide ion:

- Chlorine displaces bromide and iodide.
- Bromine displaces only iodide.
- Iodine does not displace either chloride or bromide.

When these displacement reactions happen, there are colour changes. For example, when chlorine water is mixed with potassium bromide solution, the solution changes from colourless to orange, since the chlorine has oxidised the bromide ions to bromine.

$$Cl_2(aq) + 2Br^-(aq) \longrightarrow 2Cl^-(aq) + Br_2(aq)$$

Oxidation numbers         0        −1        −1        0

When chlorine water is mixed with potassium iodide solution the solution changes from colourless to brown, since the chlorine has oxidised the iodide ions to iodine.

$$Cl_2(aq) + 2I^-(aq) \longrightarrow 2Cl^-(aq) + I_2(aq)$$

Oxidation numbers         0        −1        −1        0

# Test for halide ions

The test for halides is the silver nitrate test. This test has to be done in solution. If you start from a solid, it must first be dissolved in water.

A few drops of nitric acid are added first to make sure that any other anions (especially carbonate) are removed, as they would also form precipitates. Silver nitrate solution is then added to give:

| Ion present | Observation |
|---|---|
| $Cl^-$ | White precipitate |
| $Br^-$ | Cream precipitate |
| $I^-$ | Pale yellow precipitate |

The precipitate is the insoluble silver halide:

e.g.      $Ag^+(aq) + Cl^-(aq) \longrightarrow AgCl(s)$

Sometimes it can be difficult to tell the difference between the precipitates, especially in a single test where only one precipitate is seen. To distinguish between them, aqueous ammonia is added to the precipitate.

▲ Halide ion test results. From left to right: $I^-$, $Br^-$ and $Cl^-$ precipitate

| Precipitate | Observation |
|---|---|
| AgCl | Precipitate dissolves in dilute $NH_3$ |
| AgBr | Precipitate does not dissolve much in dilute $NH_3$ but does dissolve in conc. $NH_3$ |
| AgI | Precipitate insoluble in dilute and conc. $NH_3$ |

**Exam tip**

Halogen should be used when describing the atom or molecule. Halide should be used when describing the negative ion, e.g. $Cl_2$ is chlorine but $Cl^-$ is chloride.

**Knowledge check** 7

Describe what happens when bromine water is added to potassium iodide solution. Explain why this reaction is classified as a redox reaction.

## Uses of chlorine and fluoride in water treatment

Chlorine is commonly added to water as the gaseous element and the following equilibrium is established:

$$Cl_2 + H_2O \rightleftharpoons HCl + HOCl$$

The chlorate ion, $ClO^-$, kills bacteria and other microbes, so adding chlorine to water makes it safe to drink (or swim in). In particular, chlorination is used to prevent the outbreak of serious diseases such as typhoid and cholera.

However, there are risks in using chlorine to treat water. It is highly toxic and it can react with naturally occurring organic compounds found in the water supply to form chlorinated hydrocarbons which can cause liver and kidney cancer.

The risks are small compared with the risks from untreated water and there appear to be only beneficial effects below one part per million (1 ppm). However, some people object to water chlorination as forced 'mass medication'.

Fluoride is generally added to water to reduce tooth decay by preventing cavities. Water fluoridation reduces cavities in children but its effectiveness in adults is less clear. Although fluoridation can cause dental fluorosis, which leads to tooth discolouration, there is no clear evidence of other adverse effects from water fluoridation. It appears to have only beneficial effects below 1 ppm.

However, many people object to water fluoridation as forced 'mass medication'. Given the prevalence of fluoride in toothpaste, mouth rinses and other dental products, many people think that adding fluoride to water supplies (or bottled water) can be detrimental to long-term dental health.

**Knowledge check**

State one benefit and one risk of adding: a) chlorine and b) fluoride to drinking water.

# Practical activity

## Soluble salt formation

For example, copper(II) sulfate can be formed by neutralising sulfuric acid with the insoluble base copper(II) oxide,

$$H_2SO_4(aq) + CuO(s) \longrightarrow CuSO_4(aq) + H_2O(l)$$

These are the steps in the formation of the salt.

1. Some copper(II) oxide is added to dilute sulfuric acid. More is added until no more dissolves. (Warming might be necessary.) The solution turns blue.
2. All the acid has been used up. The excess solid is removed by filtering. This leaves a blue solution of copper(II) sulfate in water.
3. The solution is heated to evaporate some of the water.
4. It is left to cool. Blue crystals of copper(II) sulfate start to form.

The water should not be fully evaporated because if this happens, a powder will form rather than crystals.

If copper(II) carbonate is used, the method is exactly the same but effervescence (fizzing) is seen when the carbonate is added to the acid because carbon dioxide is given off. When no more effervescence is seen, all the acid has been used up.

▲ Step 1: Adding CuO

▲ Step 2: Forming CuSO₄

▲ Step 3: Evaporating CuSO₄

▲ Step 4: CuSO₄ crystals

# Insoluble salt formation

An insoluble salt can be made using a precipitation reaction by reacting two suitable solutions. In a precipitation reaction the positive ions and negative ions in the two solutions switch partners to form two new compounds – one soluble salt and one insoluble salt.

Calcium carbonate, for example, can be formed from calcium nitrate and sodium chloride:

$$Ca(NO_3)_2(aq) + Na_2CO_3(aq) \longrightarrow CaCO_3(s) + 2NaNO_3(aq)$$

These are the steps in the formation of the salt:

1. Separately dissolve sodium carbonate and calcium nitrate in water and mix together using a stirring rod in a beaker.

2. Filter to remove precipitate from mixture.

3. Wash precipitate with water to remove traces of other solutions.

4. Leave in an oven to dry.

# Gravimetric analysis

Gravimetric analysis is a technique through which the amount of an analyte (the ion being analysed) can be determined through the measurement of mass. It depends on comparing the masses of two compounds containing the analyte. The mass of an ion in a pure compound can be determined and then used to find the mass percent of the same ion in a known quantity of an impure compound.

For example, the determination of chloride in a compound. Since silver chloride is insoluble and can be formed pure and is easily filtered, a soluble silver salt can be used to determine the percentage of chloride.

**‹ Link ›**

Solubility in water page 69

**≪ Practical check**

Gravimetric analysis is a **specified practical task**

In step 3 in the example, precipitation is complete when no more solid or cloudiness is formed.

In step 5, the precipitate is washed to remove any **soluble** impurities.

In step 6, the precipitate is dried to constant mass to ensure that all the water has been removed.

These are the steps in the determination of the percentage of chloride in a compound:

1. Accurately weigh a sample of the compound into a weighing bottle (e.g. 0.927 g)
2. Transfer all of the solid into a beaker. (Wash out the weighing bottle so that all the solid goes into the beaker.) Add water and stir until all the solid dissolves.
3. Add aqueous silver nitrate to the solution and allow the precipitate to settle. Test for complete precipitation by adding more drops of aqueous silver nitrate to the clear solution.
4. When no further precipitate forms, filter off the precipitate making sure that all of it has been transferred to the filter.
5. Wash the precipitate with distilled water.
6. Dry and weigh the precipitate. (e.g. 0.725 g)

## Calculation of percentage chloride

Mass of AgCl precipitate = 0.725 g

Mol of AgCl = $0.725/143.5 = 5.05 \times 10^{-3}$

One mole of AgCl contains one mole of Cl

Mass of Cl = $5.05 \times 10^{-3} \times 35.5 = 0.179$ g

Mass of compound = 0.927 g

% Cl = $0.179/0.927 \times 100\% = 19.3\%$

# Test yourself

1. State why lithium is described as an s-block element. [1]

2. The melting temperatures of the elements in period 3 of the periodic table are shown in the table below:

| Element | Na | Mg | Al | Si | P | S | Cl | Ar |
|---|---|---|---|---|---|---|---|---|
| Melting temperature / K | 371 | 922 | 933 | 1683 | 317 | 392 | 172 | 84 |

   (a) State which of the elements shown will not be solids at room temperature. [1]

   (b) Explain, in terms of bonding and structure, why silicon, Si, has the highest melting temperature. [2]

   (c) Explain why argon, Ar, has the lowest melting temperature. [2]

   (d) Explain why the melting temperature increases from sodium, Na, to aluminium, Al, although all three elements involve metallic bonding. [2]

   (e) Fluorine, F, is in the same group as chlorine, Cl, but is in period 2. It has a melting temperature of 53 K. Explain why chlorine's melting temperature is higher. [2]

3. Explain why the following reaction can be classified as a redox reaction:

   $$3PbO_2 + 2Sb + 2NaOH \longrightarrow 3PbO + 2NaSbO_2 + H_2O$$ [2]

4. Frances is given four solutions:

barium chloride    magnesium chloride    potassium carbonate    sodium sulfate

She adds each of the four solutions to the others and records her results in a table.

| | barium chloride | magnesium chloride | potassium carbonate | sodium sulfate |
|---|---|---|---|---|
| barium chloride | ■ | no visible change | white precipitate | |
| magnesium chloride | | ■ | | |
| potassium carbonate | | | ■ | |
| sodium sulfate | | | | ■ |

(a) Complete the table. [4]

(b) State what she would see if she carried out a flame test on all the solutions. [1]

(c) Name the white precipitate formed when barium chloride is mixed with potassium carbonate. Write an ionic equation for its formation. [2]

(d) State what she would see if sodium hydroxide solution were added separately to the four solutions. [1]

5. A solution of strontium chloride can be identified using separate tests for strontium ions and chloride ions.

(a) A flame test can be used to prove that the solution contains strontium ions.

Briefly describe how you would carry out a flame test and state the flame colour that would be seen. [3]

(b) Give a chemical test to show that the solution contains chloride ions. Your answer should include the reagent(s) and expected observation(s). [2]

6. Halogens react directly with most metals to form a halide.

(a) Sodium burns in chlorine gas to form sodium chloride. State what you would see in this reaction. [2]

(b) Iron wool burns in chlorine gas to form iron(III) chloride but when it burns in iodine vapour iron(II) iodide is formed.

(i) Write an equation for the reaction between iron and chlorine. [1]

(ii) Explain why iron(II) iodide is formed and not iron(III) iodide. [1]

## 1.7

# Simple equilibria and acid–base reactions

So far, in this book, chemical reactions have been written as reactants $\longrightarrow$ products. However, chemical reactions do not only move in the forward direction from left to right in a chemical equation. Some reactions move in the backward direction, from right to left. This topic studies the relationship between the forward and backward reactions and their overall effect on the yield of reaction.

The word acid comes from the Latin 'acerbus' meaning sour. However, not all sour-tasting substances are acids, and very few acids are safe to taste to see if they are sour! In this topic you will learn about acids, their reactions, how to measure their acidity and how to neutralise them.

## Topic contents

You should be able to demonstrate and apply knowledge and understanding of:

- What is meant by a reversible reaction and dynamic equilibrium.
- Le Chatelier's principle in deducing the effects of changes in temperature, concentration and pressure.
- The equilibrium constant ($K_c$) and calculations involving given equilibrium concentrations.
- Acids as donors of $H^+(aq)$ and bases as acceptors of $H^+(aq)$.
- The relationship between pH and $H^+(aq)$ ion concentration $pH = -\log[H^+(aq)]$.
- Acid–base titrations.
- The difference between strong acids and weak acids in terms of relative dissociation.

### Maths skills »

- Substitute numerical values into algebraic equations.
- Change the subject of an equation.
- Use logarithmic functions on calculators.
- Use appropriate units in calculations.
- Use standard and ordinary form, decimal places and significant figures.
- Select values and find arithmetic means.
- Identify uncertainties in measurements.

# Reversible reactions

Not all chemical reactions 'go to completion', i.e. the reactants change completely to form products. Reactions do not only move in the forward direction, some reactions also move in the backward direction and products change back into reactants. These reactions are called **reversible** and the symbol '$\rightleftharpoons$' is used in equations.

You will already be familiar with a number of reversible reactions at home, in the laboratory and in industry. If you lower the temperature of water below 0 °C it freezes to ice. If the ice is allowed to reach room temperature it melts back to water. The process can be represented as:

$$\text{Water} \rightleftharpoons \text{Ice} \quad \text{or} \quad H_2O(l) \rightleftharpoons H_2O(s)$$

Blue copper(II) sulfate crystals have the formula $CuSO_4.5H_2O$ (the water is called water of crystallisation). When the copper(II) sulfate is heated, the water of crystallisation is given off as steam, leaving a white powder called anhydrous copper(II) sulfate. When water is added to the white powder, the powder gets hot and turns blue. The process can be represented as:

$$\underset{\text{blue}}{CuSO_4.5H_2O} \rightleftharpoons \underset{\text{white}}{CuSO_4} + 5H_2O$$

The Haber process, for the formation of ammonia from nitrogen and hydrogen, is a well-known industrial example:

$$N_2(g) + 3H_2(g) \rightleftharpoons 2NH_3(g)$$

**Key terms**

A **reversible reaction** is one that can go in either direction depending on the conditions.

**Dynamic equilibrium** is when the forward and reverse reactions occur at the same rate.

# Dynamic equilibria

Equilibrium is a term used to denote balance. The two main types encountered in everyday life are static and **dynamic equilibrium**. Imagine you are looking at a classroom in school. All of the class's 30 seats are being used. The students are sitting a test and half an hour later, you see that the same pupils are sitting at the same tables, so the situation has not changed, i.e. it is static. You go into another classroom, again all of the class's 30 seats are being used, but this time the students are doing group work. After a while some of the students move groups but all the seats are still occupied. This situation continues for the rest of the lesson. There is a balance between the number of students leaving a table and arriving. Although the number of students has not changed there is constant motion, i.e. it is a dynamic equilibrium.

An example of dynamic equilibrium is dissolving an ionic compound in water. When copper(II) sulfate crystals are added to water, the crystals begin to dissolve and the solution turns blue. The more copper(II) sulfate added, the deeper the blue colour. When no more dissolves and copper(II) sulfate crystals remain in the solution, the solution has become saturated and the intensity of the blue colour remains constant.

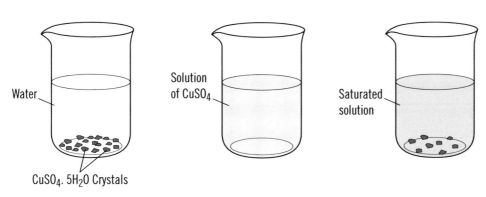

Water

$CuSO_4. 5H_2O$ Crystals

Solution of $CuSO_4$

Saturated solution

At this point the solution is in equilibrium with the undissolved solid. Although nothing seems to be happening, since the concentration of the saturated solution remains the same, the copper(II) sulfate is still dissolving but as it does so copper(II) sulfate is recrystallising from solution at the same rate.

This can be proved by using radioactive tracers. If crystals containing the radioactive $^{35}S$ are added to the saturated solution, after a time it is seen that the radioactivity is divided between the solution and the undissolved crystals.

Dynamic equilibria can also develop during chemical changes. The balance of reactants and products at equilibrium is called the equilibrium mixture. If the equilibrium mixture contains mostly products and hardly any reactants, we say that the reaction has gone to completion, (e.g. the burning of magnesium in air). On the other hand, if the equilibrium mixture consists mostly of reactants and hardly any products, we say that the reaction does not happen under these conditions, (e.g. in water vapour at room temperature, it is difficult to detect any hydrogen or oxygen present).

At equilibrium there is no observable change; the properties that we can see or measure (e.g. concentrations of reactants and products) remain constant. These are called macroscopic properties. However, the system is in constant motion – the dynamic change happening at a molecular level. For this to happen, the reactants and products must be in contact at all times so a closed system is needed, i.e. one in which substances cannot leave and cannot enter.

When equilibrium is established in a chemical process, normally the concentrations of the reactants and products change rapidly at first and then reach steady values. At this point, the reactants are being converted into products at exactly the same rate as products are being converted into reactants.

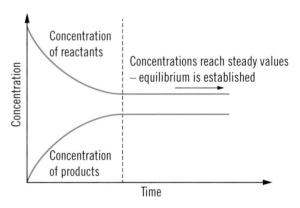

## ⟨ Link ⟩

Radioactivity uses page 22

## ≫ Study point

Features of an equilibrium are:

- There is a closed system.
- It is dynamic at a molecular level.
- The forward and reverse reactions occur at the same rate.
- Macroscopic properties remain constant.

## Exam tip ≫

If you are asked to explain what is meant by 'dynamic equilibrium' you must state that the rate of the forward reaction = the rate of the back reaction.

# Position of equilibrium

If we change the conditions of a chemical reaction in equilibrium, the equilibrium is also changed. For example, when a change in the concentration of one substance is made, the concentration of all the other substances involved in the equilibrium will change, i.e. the concentration of the reactants and products will change. The proportion of products to reactants in an equilibrium mixture is known as the **position of equilibrium**.

The response of equilibrium systems to changes of concentration, pressure and temperature were observed by Henri Le Chatelier and in 1884 he presented a principle which has important implications for many industrial processes.

**Le Chatelier's principle** states that:

If a system at equilibrium is subjected to a change then the position of equilibrium will shift to minimise that change.

## ≫ Key terms

**Position of equilibrium** is the proportion of products to reactants in an equilibrium mixture.

**Le Chatelier's principle** states that if a system at equilibrium is subjected to a change, the equilibrium tends to shift so as to minimise the effect of the change.

# Effect of concentration change

Dissolving hydrated copper(II) sulfate crystals in water gives a blue solution because the ion $[Cu(H_2O)_6]^{2+}$ is formed. When concentrated hydrochloric acid is added to a solution of copper(II) sulfate the following equilibrium is established:

$$[Cu(H_2O)_6]^{2+}(aq) + 4Cl^-(aq) \rightleftharpoons [CuCl_4]^{2-}(aq) + 6H_2O(l)$$
pale blue                              yellow–green

Adding more hydrochloric acid to the solution adds chloride ions, so the system will try to minimise this effect by decreasing the concentration of chloride ions and so the position of equilibrium will move to the right and form more $CuCl_4^{2-}$ ions making the solution yellow–green.

In the same way, addition of water to the reaction mixture moves the position of equilibrium to the left, making the solution blue.

## Study point

If the concentration of a reactant is increased, the position of equilibrium moves to the right and more products are formed.

Increasing the pressure moves the position of equilibrium to whichever side of the equation has fewer gas molecules.

An increase in temperature moves the position of equilibrium in the endothermic direction.

# Effect of pressure change

Pressure has virtually no effect on the chemistry of solids and liquids. However, it has significant effects on the chemistry of reacting gases.

The pressure of a gas depends on the number of molecules in a given volume of gas. The greater the number of molecules, the greater the number of collisions per unit time, therefore the greater the pressure of the gas.

Nitrogen dioxide, $NO_2$, a red-brown gas, is a major atmospheric pollutant from car exhausts. When two molecules of nitrogen dioxide join together to form one molecule of the colourless dinitrogen tetroxide, $N_2O_4$, the following equilibrium exists:

$$2NO_2(g) \rightleftharpoons N_2O_4(g)$$
brown            colourless

Since there are two moles of gas on the L.H.S. and one mole of gas on the R.H.S., the L.H.S. is the side at the higher pressure. If the total pressure is increased, the equilibrium will shift to minimise the increase. The pressure will decrease if the equilibrium system contains fewer gas molecules. Therefore the position of equilibrium moves to the right, which increases the yield of $N_2O_4$ and the colour of the mixture becomes lighter.

Conversely, reducing the pressure shifts the position of equilibrium to the left and the colour becomes darker.

## Exam tip

If a question asks what you would observe when an equilibrium is affected by a change in conditions, you will be expected to use the information given and state any colour changes that would occur.

## Stretch & challenge

Why does Le Chatelier's principle have important implications for many industrial processes?

# Effect of temperature change

An endothermic reaction absorbs heat from the surroundings, whereas an exothermic reaction gives out heat to the surroundings. For a reversible reaction, if the forward direction is exothermic, the backward direction is endothermic and vice versa. We can identify if a reaction is exothermic or endothermic by looking at the value of the enthalpy change, $\Delta H$. If $\Delta H$ has a negative value the reaction is exothermic, if $\Delta H$ is positive the reaction is endothermic. The enthalpy change of the forward reaction will have the same magnitude as, but the opposite sign to, the backward reaction.

Again, consider the equilibrium

$$2NO_2(g) \rightleftharpoons N_2O_4(g) \qquad\qquad \Delta H = -24\,kJ\,mol^{-1}$$
brown            colourless

Since the enthalpy change is negative, the forward reaction is exothermic. If the temperature is increased the system will try and minimise this increase. The system opposes the change by taking in heat so the position of equilibrium moves in the

## Knowledge check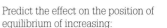

Predict the effect on the position of equilibrium of increasing:

(a) the pressure

(b) the temperature
in the following equilibrium system.

$$H_2(g) + I_2(g) \rightleftharpoons 2HI(g).$$
$\Delta H = 52\,kJ\,mol^{-1}$
Explain your answer.

 Link

Enthalpy changes page 97

endothermic direction. Therefore the equilibrium moves to the left, decreasing the yield of $N_2O_4$ and increasing the yield of $NO_2$ making the mixture appear brown.

In the same way, decreasing the temperature shifts the equilibrium to the right favouring the exothermic direction, increasing the yield of $N_2O_4$ so the mixture appears a much lighter colour.

## Effect of catalysts

A catalyst speeds up a chemical reaction by lowering the activation energy of the reaction. In a reversible reaction a catalyst will increase the rate of the forward and backward reaction to the same extent. Therefore a catalyst does not affect the position of equilibrium, but equilibrium is reached faster.

# Equilibrium constant

As we have seen in this chapter, in a reversible reaction the position of equilibrium may lie to the right, i.e. more products; or to the left, i.e. more reactants. However, the equilibrium position changes when the temperature, pressure and concentrations change. The equilibrium position may be described in precise terms by combining the equilibrium concentrations to give a value for an **equilibrium constant**. It is given the symbol $\boldsymbol{K_c}$ where the subscript $_c$ indicates that it is a ratio of concentrations.

In general for an equilibrium: $\qquad aA + bB \rightleftharpoons cC + dD$

$$K_c = \frac{[C]^c \, [D]^d}{[A]^a \, [B]^b} \qquad \text{where [C] represents the concentration of C at equilibrium, in } mol\,dm^{-3}.$$

Note that:

- The products are put in the numerator (top line) and the reactants in the denominator (bottom line).
- The concentrations are raised to powers corresponding to the mole ratio in the equation.
- The unit of $K_c$ can vary, it depends on the equilibrium.

An example is the reaction between ethanoic acid and ethanol:

$$CH_3CO_2H(aq) + C_2H_5OH(aq) \rightleftharpoons CH_3CO_2C_2H_5(aq) + H_2O(l)$$

The equilibrium constant $K_c$ is given by:

$$K_c = \frac{[CH_3CO_2C_2H_5][H_2O]}{[CH_3CO_2H][C_2H_5OH]}$$

Where $[CH_3CO_2H]$ is the equilibrium concentration of ethanoic acid in $mol\,dm^{-3}$

and the units are: $\dfrac{mol\,dm^{-3} \times mol\,dm^{-3}}{mol\,dm^{-3} \times mol\,dm^{-3}}$

therefore the units 'cancel out' and $K_c$ has no units.

However for the equilibrium

$$2NO_2(g) \rightleftharpoons N_2O_4(g)$$

$$K_c = \frac{[N_2O_4]}{[NO_2]^2}$$

and the units are: $\dfrac{mol\,dm^{-3}}{mol\,dm^{-3} \times mol\,dm^{-3}} = dm^3\,mol^{-1}$

## Study point

The equilibrium constant is not the same as the position of equilibrium.

The expression for the equilibrium constant, $K_c$, has $\dfrac{\text{products}}{\text{reactants}}$.

$K_c$ has a constant value at constant temperature because the concentrations in the equilibrium expression are raised to the powers corresponding to the mole ratios in the equation.

## 2 Knowledge check

For the following reactions write an expression for the equilibrium constant, $K_c$, giving its units.

(a) $2SO_2(g) + O_2(g) \rightleftharpoons 2SO_3(g)$

(b) $PCl_5(g) \rightleftharpoons PCl_3(g) + Cl_2(g)$

A value greater than 1 for $K_c$ shows that there are more products than reactants in the equilibrium mixture, i.e. the position of equilibrium lies to the right. The greater the value of $K_c$, the further the equilibrium lies to the right.

A value of less than 1 for $K_c$ shows that there are more reactants than products in the equilibrium mixture, i.e. the position of equilibrium lies to the left. The smaller the value of $K_c$, the further the equilibrium lies to the left.

The value of $K_c$ is constant for a particular equilibrium reaction at a constant temperature. Therefore only a change in temperature can change the value of $K_c$. Although changing the concentration or pressure can cause a shift in the position of equilibrium, it does not change the value of $K_c$.

$K_c$ has been shown to be constant for a particular reaction by setting up several experiments in which the initial amounts of reactants are varied and the equilibrium concentration of reactants and products are measured (at constant temperature).

## Worked example

For the system:

$$2H_2S(g) \rightleftharpoons 2H_2(g) + S_2(g)$$

equilibrium was reached at a temperature of 1400 K.

The equilibrium mixture contained the following concentrations:

$[H_2S] = 4.84 \times 10^{-3}\,mol\,dm^{-3}$, $[H_2] = 1.51 \times 10^{-3}\,mol\,dm^{-3}$, $[S_2] = 2.33 \times 10^{-3}\,mol\,dm^{-3}$.

Calculate the value of the equilibrium constant, $K_c$, at this temperature and give its units.

$$K_c = \frac{[H_2]^2[S_2]}{[H_2S]^2}$$

$$K_c = \frac{(1.51 \times 10^{-3})^2 \times (2.33 \times 10^{-3})}{(4.84 \times 10^{-3})^2}$$

$$K_c = 2.27 \times 10^{-4}\,mol\,dm^{-3}$$

**Knowledge check 3**

Hydrogen and iodine are reacted together to form hydrogen iodide and allowed to reach equilibrium. The equilibrium concentrations are as follows
$[H_2] = 0.11\,mol\,dm^{-3}$,
$[I_2] = 0.11\,mol\,dm^{-3}$,
$[HI] = 0.78\,mol\,dm^{-3}$.
Calculate the value of the equilibrium constant $K_c$.

## Stretch & challenge

The results for three experiments (carried out at constant temperature) for the synthesis of ammonia

$$N_2(g) + 3H_2(g) \rightleftharpoons 2NH_3(g)$$

are given in the table:

| Exp. | Initial concentration (mol dm⁻³) | | | Equilibrium concentration (mol dm⁻³) | | |
|------|------|------|------|------|------|------|
| | $N_2$ | $H_2$ | $NH_3$ | $N_2$ | $H_2$ | $NH_3$ |
| 1 | 1 | 1 | 0 | 0.922 | 0.763 | 0.157 |
| 2 | 0 | 0 | 1 | 0.399 | 1.197 | 0.203 |
| 3 | 2 | 1 | 3 | 2.59 | 2.77 | 1.82 |

Show that the value of the equilibrium constant, $K_c$, is unaffected by changes in concentration.

# The nature of acids and bases

## Acids and bases

**Acids** and **bases** make up some of the most familiar chemicals in our everyday lives, as well as being some of the most important chemicals in laboratories and industries. For these chemicals to act alike there must be some common properties in the chemicals in each of these groups.

The ion common to all acids is the hydrogen ion, $H^+$. An acid is a compound that donates $H^+$ ions (protons) in aqueous solution. This process is called dissociating. Some common acids and their formulae are:

| | |
|---|---|
| Hydrochloric acid | HCl |
| Sulfuric acid | $H_2SO_4$ |
| Nitric acid | $HNO_3$ |
| Ethanoic acid | $CH_3COOH$ |

**Study point**

Bases are normally metal oxides and hydroxides.

An alkali is a soluble base.

The equation for hydrochloric acid dissociating is given by:

$$HCl(g) \xrightarrow{water} H^+(aq) + Cl^-(aq)$$

A base is a compound that accepts $H^+$ ions from an acid. Some common bases and their formulae are:

| | | | |
|---|---|---|---|
| Magnesium oxide | MgO | Calcium oxide | CaO |
| Sodium hydroxide | NaOH | Ammonia | $NH_3$ |

If the base dissolves in water it is called an **alkali**. The ion common to all alkalis is the hydroxide ion, $OH^-$.

$$NaOH(s) \xrightarrow{water} Na^+(aq) + OH^-(aq)$$

**Stretch & challenge**

The hydrogen ion, $H^+$, is simply a proton. Since it has an extremely small diameter ($10^{-15}$ m) compared with other cations (around $10^{-10}$ m) it has a very large charge density. In aqueous solution it attracts a lone pair of electrons on a neighbouring water molecule to form a co-ordinate bond. The aqueous hydrogen ion, $H^+$(aq) actually exists as the $H_3O^+$ ion, called the oxonium ion. It has a roughly pyramidal shape.

## Strong and weak acids

We know that different acids have different strengths. It is possible to taste an acid like citric acid quite safely but sulfuric acid will quickly damage your tongue!

Since acids donate $H^+$ ions in aqueous solution, the more easily an acid can donate $H^+$ the stronger the acid.

**Link**

Formulae of common compounds page 11

The general equation for dissociation of an acid is given by:

$$HA(aq) \rightleftharpoons H^+(aq) + A^-(aq)$$
(A⁻ represents an anion)

For HCl the equilibrium lies far to the right so the equation is written as:

$$HCl(aq) \longrightarrow H^+(aq) + Cl^-(aq)$$

The acid is fully dissociated or ionised and it is described as a strong acid.

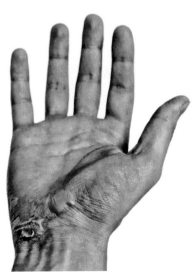

▲ Concentrated acids can burn

Therefore, a **strong acid** is fully dissociated in aqueous solution. For a monobasic acid (an acid that has one replaceable hydrogen ion such as HCl) the hydrogen ion concentration is equal to the concentration of the acid. For a dibasic acid (an acid that has two replaceable hydrogen ions such as $H_2SO_4$) the hydrogen ion concentration is double the concentration of the acid.

Many acids are far from fully dissociated in aqueous solution and these are described as weak acids, e.g. for ethanoic acid, the equilibrium

$$CH_3CO_2H(aq) \rightleftharpoons CH_3CO_2^-(aq) + H^+(aq)$$

lies to the left. In fact only about four in every thousand ethanoic acid molecules are dissociated into ions.

A **weak acid** is only partially dissociated in aqueous solution. The aqueous hydrogen ion concentration is smaller in magnitude than the concentration of the acid.

The words strong and weak only refer to the extent of dissociation and not in any way to concentration. A **concentrated acid** consists of a large quantity of acid and a small quantity of water. A **dilute acid** contains a large quantity of water.

So it is possible to have a dilute solution of a strong acid, e.g. HCl of concentration $0.0001\,mol\,dm^{-3}$ or a concentrated solution of a weak acid, e.g. $CH_3CO_2H$ of concentration $8\,mol\,dm^{-3}$.

Similarly bases can be classified as strong or weak. An example of a strong base is NaOH and an example of a weak base is $NH_3$.

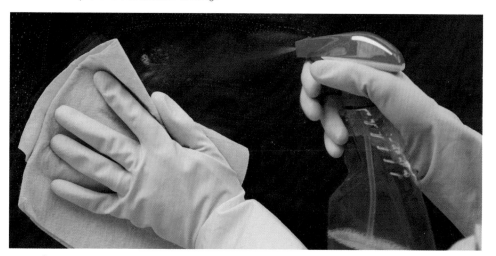

▲ Dilute acids can be used to clean glass

# The pH scale

Since acids are compounds that donate $H^+$ ions (aq), the acidity of a solution is a measure of the concentration of the aqueous hydrogen ion, $H^+$(aq).

However, $H^+$(aq) concentration varies over a wide range and can be extremely small, e.g. from $1 \times 10^{-14}$ to $1\,mol\,dm^{-3}$. This wide variation made it very difficult for people to deal with the concept of acidity.

This problem was overcome in 1909 when the Danish chemist Soren Sorenson proposed the pH scale (p stands for 'Potenz', the German for strength). He defined pH as:

$$pH = -\log_{10}[H^+] \qquad \text{where } [H^+] \text{ is the concentration of } H^+ \text{ in } mol\,dm^{-3}$$

The negative sign in the equation results in pH decreasing as the aqueous hydrogen ion concentration increases.

>> **Key terms**

A **strong acid** is one that fully dissociates in aqueous solution.

A **weak acid** is one that partially dissociates in aqueous solution.

>> **Study point**

This is not strictly true for $H_2SO_4$ because the dissociation occurs in two stages and the dissociation in the second stage is incomplete. However, at A Level, 100% dissociation is assumed.

**Knowledge check** 4

Explain why nitric acid is classified as a strong acid.

**Knowledge check** 5

Differentiate clearly between a strong acid and a concentrated acid.

**Study point**

The simplest way to measure pH is by using universal indicator. More accurate measurements can be made by using pH meters.

 **Study point**

The higher the $H^+$ ion concentration, the lower the pH and the stronger the acid.

**Worked examples**

1. What is the pH of:

   (a) a sample of rain water with a $H^+(aq)$ concentration of $3.9 \times 10^{-6}\,mol\,dm^{-3}$?

   $$pH = -\log(3.9 \times 10^{-6})$$
   $$pH = 5.4$$

   (b) A solution with a $H^+(aq)$ concentration of $1.0\,mol\,dm^{-3}$?

   $$pH = -\log 1$$
   $$pH = 0$$

2. Sometimes it is useful to determine aqueous hydrogen ion concentrations from pH values, e.g. a sample of acid rain has a pH of 2.2. What is the aqueous hydrogen ion concentration of this sample?

   $$pH = -\log[H^+]$$

multiply both sides by $-1$ and rearrange

   $$\log[H^+] = -pH$$
   $$\log[H^+] = -2.2$$

take the antilogarithm of both sides (often 'shift log' on a calculator)

   $$[H^+] = 10^{-2.2}$$
   $$[H^+] = 6.3 \times 10^{-3}\,mol\,dm^{-3}$$

Using the pH scale the acidity of any solution can be expressed as a simple more manageable number, ranging from 0 to 14. This is much more convenient for the general public when dealing with concepts of acidity.

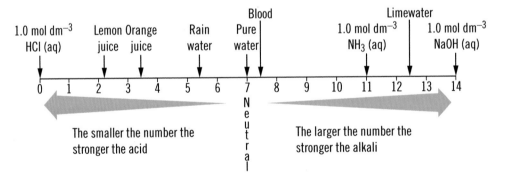

 **Key term**

A **salt** is the compound that forms when a metal ion replaces the hydrogen ion in an acid.

 **Study point**

Salts of HCl are chlorides.
Salts of $H_2SO_4$ are sulfates.
Salts of $HNO_3$ are nitrates.

# Neutralisation

The reactions of acids with bases and carbonates are not specifically mentioned in the specification. However, since all learners are expected to demonstrate knowledge and understanding of standard content covered at GCSE level, this section gives a recap on the minimum knowledge that is required about these reactions.

Bases (including alkalis) and carbonates react with acids in neutralisation reactions. The base (or carbonate) accepts the $H^+$ ions donated by the acid and the $H^+$ ions of the acid are replaced by metal ions (or $NH_4^+$ ions) to form a **salt**. For example, aqueous hydrochloric acid and aqueous sodium hydroxide react according to the equation:

$$HCl(aq) + NaOH(aq) \longrightarrow NaCl(aq) + H_2O(l)$$

If we write an ionic equation, we get:

$$H^+(aq) + Cl^-(aq) + Na^+(aq) + OH^-(aq) \longrightarrow Na^+(aq) + Cl^-(aq) + H_2O(l)$$

Removing the spectator ions gives:

$$H^+(aq) + OH^-(aq) \longrightarrow H_2O(l)$$

If an insoluble metal oxide, e.g. MgO, reacts with an acid, e.g. $H_2SO_4$, we get:

$$MgO(s) + H_2SO_4(aq) \longrightarrow MgSO_4(aq) + H_2O(l)$$

The oxide ion, $O^{2-}$, has accepted the $H^+$ ion to form water and the salt magnesium sulfate is formed.

If a carbonate, e.g. $PbCO_3$, reacts with an acid, e.g. $HNO_3$, we get:

$$PbCO_3(s) + 2HNO_3(aq) \longrightarrow Pb(NO_3)_2(aq) + H_2O(l) + CO_2(g)$$

This time the carbonate ion has accepted the $H^+$ ion to form water and carbon dioxide and the salt lead nitrate is formed.

Neutralisation always produces a salt. Many salts are very useful and neutralising an acid is a convenient way to form salts.

The same method can be used to form a salt from an insoluble base or carbonate (for details see pages 72–73) but a different method has to be used to form a salt from an alkali since alkalis are soluble in water.

The method for preparing a salt from an acid and an alkali (or a soluble carbonate) is known as titration. (Details are shown on pages 86–87.)

A volume of alkali is measured into a flask and a few drops of indicator are added. Acid is added from a burette until the indicator changes colour. When the volume of acid needed to neutralise the alkali has been calculated, the procedure is repeated without the indicator so the correct amount of acid is added to the flask. The solution from the flask is heated to evaporate some of the water. Then it is left to cool and form crystals of pure salt.

**Knowledge check**

Name the products formed when nitric acid reacts with potassium hydroxide.

 **Study point**

The general equations for neutralisation are;

acid + base $\longrightarrow$ salt + water

acid + alkali $\longrightarrow$ salt + water

acid + carbonate $\longrightarrow$ salt + water + carbon dioxide

**Study point**

A salt can also be made by reacting a metal with an acid in a similar way to neutralising a base with an acid.

 **Practical check**

Preparing a soluble salt by titration is a **specified practical task**.

When the titration is repeated without an indicator, make sure that when nearing the exact amount needed, the acid is added a drop at a time in order to avoid overshooting.

If you evaporate all of the water then a powder will form instead of crystals.

# Practical activity

## Acid-base titrations

An acid–base titration is a type of volumetric analysis where the volume of one solution, an acid, that reacts exactly with a known volume of another solution, a base, is measured. The precise point of neutralisation is measured using an indicator. Titrations are not just used for preparing salts, they are often used for calculating the exact concentrations of acid or base solutions. To do this one of the solutions must be a **standard solution** or it must have been standardised. In the analysis you use the standard solution to find out information about the substance dissolved in the other solution.

**Key term**

A **standard solution** is one for which the concentration is accurately known.

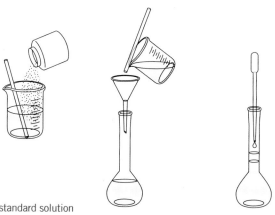

▲ Preparing a standard solution

>> **Study point**

To calculate the mass of reagent required to make a standard solution use the expressions $n = cV$ and $m = n \times M$. For example, what mass of anhydrous sodium carbonate must be dissolved in 250 cm³ of solution to obtain a concentration of 0.0500 mol dm⁻³?

$$n(Na_2CO_3) = 0.250 \times 0.0500$$
$$= 1.25 \times 10^{-2} \, mol \, dm^{-3}$$
$$\text{mass } Na_2CO_3 = 1.25 \times 10^{-2} \times 106$$
$$= 1.325 \, g.$$

**Practical check** >>

In the preparation of a standard solution:

Use the tare button on the weighing balance so that the scale reads zero.

When adding solid to the weighing bottle, remove the bottle from the pan and then add the solid, checking the mass until the correct amount has been added. This prevents errors caused by spilling solid onto the pan of the balance.

**8** **Knowledge check**

Give two reasons why sodium hydroxide is unsuitable as a primary standard to prepare a standard solution.

◀**Stretch & challenge**

Two important terms in titration are 'end point' and 'equivalence point'. The end point of a titration is when the indicator changes colour, i.e. it is a property of the indicator, it does not mean that the reaction is complete. The equivalence point of the titration occurs when the two solutions have reacted exactly, i.e. moles of acid = moles of base.

There are many different acid–base indicators and it is important to choose the correct one for a particular reaction. If the correct indicator is chosen, the end point of the titration will coincide exactly with its equivalence point. Also the indicator should show a distinct colour change.

## Standard solution

A standard solution is prepared using a primary standard. A primary standard is typically a reagent which can be weighed easily, and which is so pure that its weight is truly representative of the number of moles of substance contained. Features of a primary standard include:

1. High **purity**
2. Stability (**low reactivity**)
3. Low **hygroscopicity** (to minimise weight changes due to humidity)
4. High **molar mass** (to minimise weighing errors).

Sodium hydroxide cannot be used as a primary standard because it reacts with atmospheric carbon dioxide.

Two examples of primary standards are:

- **Potassium hydrogen phthalate** (usually called KHP) for standardisation of aqueous base solutions.
- **Sodium carbonate** for standardisation of aqueous acids.

A standard solution is prepared from a solid as follows:

- Calculate the mass of the solid required and accurately weigh this amount into a weighing bottle.
- Transfer all of the solid into a beaker. Wash out the weighing bottle so that all the weighings run into the beaker. Add water and stir until all the solid dissolves.
- Pour all the solution carefully through a funnel into a **volumetric (graduated) flask**, washing all the solution off the beaker and the glass rod. Add water until just below the graduation mark.
- Add water drop by drop until the graduation mark is reached and mix the solution thoroughly.

## Performing a titration

All titrations involve the same major equipment / chemicals, namely:

- A burette containing one solution (e.g. an acid).
- A conical flask containing the other solution (e.g. a base).
- A pipette to accurately transfer the solution to the conical flask.
- An indicator to show when the reaction is completed.

(Two common indicators are phenolphthalein which is colourless in acid solution and purple in alkaline solution, and methyl orange which is red in acid solution and yellow in alkaline solution.)

All titrations follow the same overall method:

- Pour the acid into a **burette**, using a funnel, making sure that the jet is filled. Remove the funnel and read the burette.
- Use a **pipette** to add a measured volume of the base into a **conical flask**.
- Add a few drops of indicator to the solution in the flask.
- Run the acid from the burette into the solution in the conical flask, swirling the flask.
- Stop when the indicator just changes colour (this is the end-point of the titration).
- Read the burette again and subtract to find the volume of acid used (this is known as the titre).

▲ The burette is filled with acid using a funnel

▲ A graduated pipette is used to measure 25 $cm^3$ of alkali into a conical flask

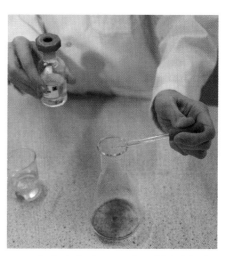

▲ An indicator, e.g. phenolphthalein, is added to the alkali which turns purple

▲ Acid is added to the alkali, swirling the flask continually

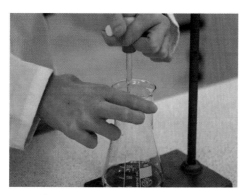

▲ The indicator has changed colour, showing that the end-point has been reached

- Repeat the titration, making sure that the acid is added drop by drop near the end – point, until there are at least two readings that are within 0.20 $cm^3$ of each other and calculate a mean titre.

## Worked example

A student is given a mixture containing sodium carbonate and an unreactive compound. She is asked to determine the percentage of sodium carbonate in the mixture.

She weighs 2.05 g of the mixture, accurately makes up the solution to 250 $cm^3$ with distilled water and titrates 25.0 $cm^3$ of this solution with hydrochloric acid of concentration 0.100 mol $dm^{-3}$. She repeats the titration three times and the average titre is 23.20 $cm^3$.

The equation for the reaction between sodium carbonate and hydrochloric acid is:

$Na_2CO_3 + 2HCl \longrightarrow 2NaCl + H_2O + CO_2$

Calculate the percentage of $Na_2CO_3$ in the mixture.

Moles HCl = $\dfrac{0.1 \times 23.20}{1000}$ = 2.32 × $10^{-3}$

From equation 1 mol $Na_2CO_3$ reacts with 2 mol HCl

Moles $Na_2CO_3$ = 1.16 × $10^{-3}$

Mass $Na_2CO_3$ = 1.16 × $10^{-3}$ × 106 = 0.123 g

Mass in original solution = 0.123 × 10 = 1.23 g

% $Na_2CO_3$ = $\dfrac{1.23}{2.05}$ = 60.0 %

### ❮❮ Practical check

Standardising an acid solution is a **specified practical task**.

A 25 $cm^3$ pipette is used rather than a 25 $cm^3$ measuring cylinder because a pipette is more precise. A pipette has an uncertainty of ±0.06 $cm^3$ while a measuring cylinder has an uncertainty of ±0.5 $cm^3$.

When using a pipette, always use a pipette filler – never suck the liquid up by mouth.

Never blow the contents of a pipette into the conical flask.

The conical flask can be wet as this will not alter the number of moles of alkali/acid added from the pipette, but the flask must not contain any acid or alkali.

The burette should be rinsed out with a few $cm^3$ of the appropriate acid/alkali.

Removing the funnel prevents some of the solution dripping into the burette, causing an incorrect reading to be recorded.

Record the volume in the burette by looking at where the bottom of the meniscus is on the scale.

Always estimate the volume to the nearest 0.05 $cm^3$.

When calculating the mean only use concordant results.

**Knowledge check**

During an acid–base titration, explain why:

(a) A measuring cylinder is not used to transfer the basic solution into a conical flask.

(b) An indicator is added to the basic solution in the conical flask.

 **Practical check**

Another example of a double titration is to determine the concentrations of sodium carbonate and sodium hydrogencarbonate in a mixture.

**Study point**

Make sure that you can explain why each step is carried out in a titration.

## Double titrations

Since different indicators change colour at different pH values, if a solution contains a mixture of two bases which are of different strengths, one titration can be performed, but in two stages, using two different indicators, one added at each stage, to calculate the concentrations of both bases.

For example, the concentrations of sodium hydroxide, NaOH, and sodium carbonate, $Na_2CO_3$, in a mixture can be determined by titrating with hydrochloric acid, HCl, using phenolphthalein and methyl orange as indicators.

Phenolphthalein changes colour at around pH 9 (pink to colourless).

Methyl orange changes colour at around pH 4 (yellow to orange).

Sodium hydroxide is neutralised by the acid according to:

$$OH^-(aq) + H^+(aq) \longrightarrow H_2O(l)$$

When sodium carbonate is neutralised by the acid two reactions occur:

$$CO_3^{2-}(aq) + H^+(aq) \longrightarrow HCO_3^-(aq)$$
$$HCO_3^-(aq) + H^+(aq) \longrightarrow H_2O(l) + CO_2(g)$$

However, at pH 9 all the hydroxide ions have been neutralised and the carbonate ions have been converted to hydrogencarbonate ions. At pH 4 the hydrogencarbonate ions are converted to water and carbon dioxide:

i.e. $\left. \begin{array}{l} OH^-(aq) + H^+(aq) \longrightarrow H_2O(l) \\ CO_3^{2-}(aq) + H^+(aq) \longrightarrow HCO_3^-(aq) \end{array} \right\}$ both completed at phenolphthalein stage

$HCO_3^-(aq) + H^+(aq) \longrightarrow H_2O(l) + CO_2(g)$      completed at methyl orange stage

Therefore the first stage of the titration (change in colour of the phenolphthalein) relates to the concentration of the hydroxide and the carbonate.

The second stage of the titration (after addition of methyl orange) relates to the concentration of carbonate only. (Since 1 mol $HCO_3^-$ = 1 mol $CO_3^{2-}$.)

### Worked example

$25.0 \, cm^3$ of a solution containing sodium hydroxide and sodium carbonate were titrated against dilute hydrochloric acid of concentration $0.100 \, mol \, dm^{-3}$ using phenolphthalein as indicator. After $22.00 \, cm^3$ of acid had been added the indicator was decolourised. Methyl orange was added and a further $8.25 \, cm^3$ of acid were needed to turn the indicator orange.

Calculate the concentrations of sodium hydroxide and sodium carbonate in the solution.

$$\text{Moles HCl (in first stage)} = \frac{22.00}{1000} \times 0.100 = 2.20 \times 10^{-3} \, mol$$

Therefore    moles $OH^- + CO_3^{2-} = 2.20 \times 10^{-3} \, mol$

$$\text{Moles HCl (in second stage)} = \frac{8.25}{1000} \times 0.100 = 8.25 \times 10^{-4} \, mol$$

Therefore      moles $HCO_3^- = 8.25 \times 10^{-4} \, mol$

Since      1 mol $HCO_3^-$ = 1 mol $CO_3^{2-}$

Moles $CO_3^{2-} = 8.25 \times 10^{-4} \, mol$

Moles $OH^- = (2.20 \times 10^{-3}) - (8.25 \times 10^{-4}) = 1.375 \times 10^{-3} \, mol$

$$\text{Concentration } CO_3^{2-} = \frac{8.25 \times 10^{-4}}{0.025} = 0.0330 \, mol \, dm^{-3}$$

$$\text{Concentration } OH^- = \frac{1.375 \times 10^{-3}}{0.025} = 0.0550 \, mol \, dm^{-3}$$

## Back titrations

Sometimes it is not possible to use standard titration methods. For example, the reaction between a certain substance and titrant can be too slow, there can be a problem with end-point determination or the base is an insoluble salt. In such situations we often use a technique called back titration.

In back titration a known excess of one reagent **A** reacts with an unknown amount of reagent **B**. At the end of the reaction, the amount of reagent **A** that remains is found by titration. A simple calculation gives the amount of reagent **A** that has been used and the amount of reagent **B** that has reacted.

**≪ Practical check**

Performing a back titration and a double titration are **specified practical tasks**.

Exact details of titrations may differ, but the procedures will remain the same.

**⟨ Link ⟩**

Acid-base titration calculations pages 42–43

### Worked example

Limestone is a sedimentary rock composed primarily of calcium carbonate. A 1.00 g sample of limestone is allowed to react with 100 cm$^3$ of dilute hydrochloric acid of concentration 0.200 mol dm$^{-3}$ for neutralisation. The excess acid required 22.8 cm$^3$ of aqueous sodium hydroxide solution of concentration 0.100 mol dm$^{-3}$.

Calculate the percentage of calcium carbonate in the limestone sample.

Equation for titration is given by:

$$NaOH(aq) + HCl(aq) \longrightarrow NaCl(aq) + H_2O(l)$$

$$\text{Moles NaOH} = \frac{22.8}{1000} \times 0.100 = 2.28 \times 10^{-3} \, mol$$

From equation 1 mol NaOH reacts with 1 mol HCl

$$\text{Moles HCl in excess} = 2.28 \times 10^{-3} \, mol$$

Equation for reaction between calcium carbonate and acid is given by:

$$CaCO_3(s) + 2HCl(aq) \longrightarrow CaCl_2(aq) + H_2O(l) + CO_2(g)$$

$$\text{Original moles HCl added} = \frac{100}{1000} \times 0.200 = 0.0200 \, mol$$

$$\text{Moles acid reacted with} = 0.0200 - (2.28 \times 10^{-3}) = 0.01772 \, mol$$
calcium carbonate

From equation 2 mol HCl reacts with 1 mol CaCO$_3$

$$\text{Moles CaCO}_3 = \frac{0.01772}{2} = 8.86 \times 10^{-3} \, mol$$

$$\text{Mass CaCO}_3 = 8.86 \times 10^{-3} \times 100 = 0.886 \, g$$

$$\% \, CaCO_3 = \frac{0.886}{1.00} \times 100 = 88.6\%$$

# Test yourself

1.  Hydrogen and chlorine react together to form hydrogen chloride:

    $$H_2(g) + Cl_2(g) \rightleftharpoons 2HCl(g) \qquad \Delta H = -184 \text{ kJ mol}^{-1}$$

    (a) State Le Chatelier's principle. [1]

    (b) State, giving your reasons, how the equilibrium yield of hydrogen chloride is affected, if at all, by:

    (i) increasing the temperature at constant pressure. [2]

    (ii) increasing the pressure at constant temperature. [2]

2.  (a) Nitrogen and hydrogen combine to form ammonia:

    $$N_2(g) + 3H_2(g) \rightleftharpoons 2NH_3(g) \qquad \Delta H = -92 \text{ kJ mol}^{-1}$$

    Complete the table which refers to the effect a change in conditions has on the position of equilibrium. [3]

| Change | Effect, if any, on position of equilibrium | Effect, if any, on value of $K_c$ |
|---|---|---|
| Increase in pressure | Shift to the right | |
| Drop in temperature | | |

    (b) One use of ammonia is in the production of nitric acid. In the first part of this process ammonia is oxidised in air:

    $$4NH_3(g) + 5O_2(g) \rightleftharpoons 4NO(g) + 6H_2O(g) \qquad \Delta H = -900 \text{ kJ mol}^{-1}$$

    State, giving your reasons, the general conditions of temperature and pressure required to give a high equilibrium yield of nitric oxide in this process. [4]

3.  Chlorine dissolves in water to produce a mixture of hydrochloric acid and chloric(I) acid. Hydrochloric acid is a strong acid but chloric(I) acid is a weak acid.

    $$Cl_2(g) + H_2O(l) \rightleftharpoons HCl(aq) + HOCl(aq)$$

    (a) Use Le Chatelier's principle to explain how the concentration of hydrogen ions, $H^+(aq)$, would change if more chlorine were dissolved in a solution that had reached dynamic equilibrium. [2]

    (b) Explain the difference between the meaning of the terms *strong acid* and *weak acid*. [2]

    (c) A 0.1 mol dm$^{-3}$ solution of chloric(I) acid has a pH of 4.3. Calculate the hydrogen ion concentration of this solution. [2]

4.  A flask containing a mixture of ethanoic acid and ethanol was maintained at 25 °C until the following equilibrium had been established:

    $$CH_3COOH + C_2H_5OH \rightleftharpoons CH_3COOC_2H_5 + H_2O$$

    (a) Write an expression for the equilibrium constant, $K_c$, and give its units if any. [2]

    (b) At this temperature, the equilibrium constant, $K_c$, has a numerical value of 4.07.

    If the equilibrium concentration of $CH_3COOC_2H_5$ and $H_2O$ are both $9.13 \times 10^{-2}$ mol dm$^{-3}$ and that of $CH_3COOH$ is $1.09 \times 10^{-1}$ mol dm$^{-3}$, calculate the equilibrium concentration of $C_2H_5OH$ at this temperature. [2]

5. Frances measures a mass of hydrated copper(II) sulfate, $CuSO_4.5H_2O$, and uses this to make exactly 250.0 cm$^3$ of copper(II) sulfate solution of concentration 0.250 mol dm$^{-3}$.

    (a) Calculate the mass of hydrated copper(II) sulfate required to prepare this solution. [2]

    (b) Describe, giving full practical details, how Frances should prepare the 250.0 cm$^3$ of copper(II) sulfate solution. [5]

6. Elinor titrated a solution of sodium hydroxide against a 0.300 mol dm$^{-3}$ hydrochloric acid solution to find the concentration of the sodium hydroxide.

    She obtained the following results using 25.0 cm$^3$ samples of the sodium hydroxide solution.

| Titration | 1 | 2 | 3 | 4 |
|---|---|---|---|---|
| Final reading (cm$^3$) | 23.45 | 23.20 | 23.10 | 23.40 |
| Initial reading (cm$^3$) | 0.00 | 0.25 | 0.20 | 0.30 |
| Titre (cm$^3$) | | | | |

    (a) Calculate the mean titre that Elinor should use in her calculations. [2]

    (b) Describe the practical steps that Elinor used to obtain a titration value. You should start by measuring the acid into a burette with 25.0 cm$^3$ of the sodium hydroxide solution already in a conical flask. [6]

    (c) Nick carried out the titration using a 1.50 mol dm$^{-3}$ hydrochloric acid solution instead of a 0.300 mol dm$^{-3}$ solution. State, and explain, if Nick is likely to have got a more accurate result than Elinor for the concentration of the sodium hydroxide solution. [2]

# Unit 1

It is really important to read examination questions very carefully. Resist the temptation to write all you know on a topic, just having spotted a particular word in a question. Be sure you know what is being asked and that you apply your knowledge to the specific examples being discussed.

Some questions will contain command words. These are words that tell you what to do. Make sure you understand command words and that you do what they say.

**‹Link›**

Command words are explained on page 8.

Many questions increase in complexity as you go through:

- A question may start by asking you to name a structure, make a statement or give the meaning of a term. Such questions require you to have learned your notes by heart.

- Next you may be asked to apply your knowledge, e.g. in a straightforward calculation, interpreting data or explaining a chemical concept.

- You may then be asked to perform a multi-stage calculation, evaluate evidence or suggest how to improve a particular experiment.

The example below shows this increase in complexity in a question covering Unit 1 topics. The command words are in **bold**.

(a) **State** the meaning of the term *molar first ionisation energy*. [1]

*Since the command word is 'state', you simply have to recall a chemical fact and write it down. As the question specifically asks for 'molar' you have to include '1 mole' in your answer.*

(b) State and **explain** how you would expect the first ionisation energy of beryllium to **compare** with the first ionisation energy of magnesium. [2]

*In this question, since there are only two possible answers, 'higher' or 'lower', there will not be a mark for the 'state' part. Both marks are for the 'explain' part, so two reasons are needed. Also, because the question specifically asks you to 'compare', you must refer to both elements or use comparative language, e.g. 'higher', 'closer'.*

*Make sure that your answer is unambiguous. Be very careful about making statements with the word 'it'. If there is any ambiguity you will not gain the mark.*

(c) The first ionisation energy of magnesium is 736 kJ mol$^{-1}$. This value can be found from the wavelength of a line in the atomic spectrum of magnesium. **Calculate** the wavelength of this line in nanometres, giving your answer to an appropriate number of significant figures. [4]

*In this question you will have to use different equations to calculate the wavelength and you will need to rearrange the equations to make wavelength the subject, i.e. the term on its own on one side of the equation. You will also have to convert from one unit of measurement to another to ensure that the answer has the correct unit. Remember to use the Data Booklet to obtain the values of any constants and to help with converting between units.*

*Always show your working clearly, even if your final answer is wrong, you could gain many of the available marks due to error carried forward. Since the calculation requires an appropriate number of significant figures, always give your answer to the lowest number of significant figures given in the question.*

1  Using ideas that you have studied in your Chemistry course, comment on and explain the following observations.

   (a)  In sodium chloride and caesium chloride the arrangements of the particles in the solids are different.  [4]

   (b)  Hydrogen sulfide, $H_2S$, is a gas at room temperature and pressure, but water, $H_2O$, is a liquid under the same conditions.  [4]

   (c)  The bond angles in the $PCl_4^+$ ion are greater than the bond angles in the $PCl_6^-$ ion.  [4]

   (Total 12 marks)

   *[Eduqas Component 1 2016 Q10]*

2  (a)  Some students were discussing ionisation energies.

   (i)  State the meaning of the term *standard molar first ionisation energy*.  [2]

   (ii)  The graph below shows the logarithm of the first eight successive ionisation energies for element X.

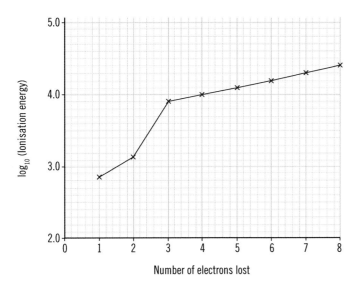

   (iii)  Explain why successive ionisation energies increase.  [2]

   (iv)  Use the graph to determine in which group of the periodic table element X is found. Explain your answer.  [2]

   (b)  The students used a hydrogen discharge tube and observed some coloured lines. One student said that these visible lines could be used to determine the value of the first ionisation energy of hydrogen.

   (i)  Discuss whether you agree with this student's suggestion. You should include an explanation of how spectral lines are produced and how they can be used to determine ionisation energy.  [6 QER]

   (ii)  A spectral line with a frequency of $3.28 \times 10^{15}$ Hz was formed when an atom of hydrogen was ionised.

   Calculate the first ionisation energy of hydrogen in kJ $mol^{-1}$. Give your answer to the appropriate number of significant figures.  [3]

   (Total 15 marks)

   *[Eduqas Component 1 2017 Q10]*

**3** (a) Caroline was investigating the number of moles of water of crystallisation, $x$, in hydrated barium chloride, $BaCl_2.xH_2O$. She was told that $x$ is a whole number.

She followed an instruction sheet:

- Weigh an empty crucible with its lid.
- Add the hydrated salt to the crucible and weigh the crucible, lid and salt.
- Place the lid on the crucible and heat salt for 3 minutes.
- Cool and reweigh the crucible, lid and contents.
- Heat for another 2 minutes and cool and reweigh again.

Caroline obtained the following results:

| | Mass/g |
|---|---|
| Crucible + lid | 13.132 |
| Crucible + lid + BaCl$_2$.xH$_2$O | 15.051 |
| Crucible + lid + contents (after 1st heating) | 14.787 |
| Crucible + lid + contents (after 2nd heating) | 14.777 |

(i) Use the data to determine the value of $x$ in the formula $BaCl_2.xH_2O$. You must show your working. [4]

(ii) Why did the instructions say that the lid should be in place when the heating was carried out? [1]

(iii) Ethan said that Caroline's method was inaccurate, even though she had carried out the experiment carefully and recorded all her results correctly.

Suggest two ways in which Caroline could make her experiment more accurate.

Explain your answers. [4]

(iv) Caroline agreed that her experiment had been inaccurate but said that it gave the correct answer for $x$. Comment on why Caroline was correct and that accuracy need not be high in this experiment to determine the value of $x$. [1]

(b) Caroline used the barium chloride as one of the reagents to identify the ions present in an aqueous solution **W**. Solution **W** contains only two ions.

The reagents were added to small volumes of solution **W** and the following observations were made:

| Test | Observation |
|---|---|
| Add aqueous sodium hydroxide | No visible reaction |
| Add aqueous barium chloride | White precipitate formed |
| Add dilute nitric acid | Vigorous effervescence seen |

(i) From these observations, name one ion present in solution **W**. [1]

(ii) The observations allowed Caroline to eliminate some metal ions as being present in **W**. Suggest one metal ion that she eliminated. [1]

(iii) Write the ionic equation for the reaction between aqueous solutions of barium chloride and **W**. Include state symbols. [1]

(Total 13 marks)

[*Eduqas Component 1 2017 Q12*]

**4** (a) Several different compounds containing sodium, chlorine and oxygen exist. One of these decomposes on heating as shown in the equation.

$$2NaClO_3(s) \longrightarrow 2NaCl(s) + 3O_2(g)$$

 (i) What is the oxidation state of chlorine in $NaClO_3$? [1]

 (ii) Calculate the maximum volume of gas (in $dm^3$), measured at 600 K and 1 atm pressure, that can be made by heating 88.0 g of $NaClO_3$. Give your answer to an appropriate number of significant figures. [3]

 (b) The active component of bleach is sodium chlorate(I), NaClO. This is prepared by passing chlorine into aqueous sodium hydroxide.

$$2NaOH(aq) + Cl_2(g) \longrightarrow NaClO(aq) + NaCl(aq) + H_2O(l)$$

 Calculate the atom economy (in %) of this process when used to prepare sodium chlorate(I). [2]

 (c) Another compound containing sodium, chlorine and oxygen has the following composition by mass.

 Na 18.8 %          Cl 29.0 %          O 52.2 %

 Calculate its empirical formula. [2]

(Total 8 marks)

*[Eduqas Component 1 2019 Q12]*

**5** Carboxylic acids react with alcohols to make esters, using sulfuric acid as a catalyst. These reactions are reversible.

$$CH_3COOH \quad + \quad C_2H_5OH \rightleftharpoons CH_3COOC_2H_5 + H_2O$$

    ethanoic acid          ethanol          ethyl ethanoate

(a) State what is meant by a *reversible* reaction. [1]

(b) In an experiment to prepare ethyl ethanoate, 3.0 mol of ethanoic acid were mixed with 2.5 mol of ethanol and a small amount of concentrated sulfuric acid. Water was added to make a total volume of 1.0 $dm^3$.

The number of moles of ethanoic acid present was measured as the reaction proceeded until equilibrium was reached. The results were then plotted.

On the grid sketch:

- the line that shows the number of moles of ethanol as the reaction proceeds to equilibrium. Label this line **A**.

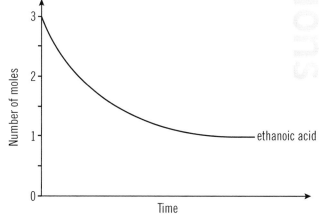

- the line that shows the number of moles of ethyl ethanoate as the reaction proceeds to equilibrium. Label this line **B**. [3]

(c) (i) Write the expression for the equilibrium constant, $K_c$, for the esterification reaction. Include the unit, if any. [2]

 (ii) Under certain conditions the value of $K_c$ was found to be 4. At a higher temperature, with all other factors being kept constant, this value remained almost the same. Explain what can be deduced from this information. [2]

(d) Esterification is catalysed by the addition of concentrated sulfuric acid. In an esterification the final pH of the solution was 2.4. Calculate the concentration of hydrogen ions present, in mol $dm^{-3}$. [2]

(e) 2.94 g of ethanoic acid were mixed with 2.07 g of ethanol and allowed to react. 2.73 g of ethyl ethanoate were produced. Calculate the percentage yield of this reaction. [4]

(Total 13 marks)

*[Eduqas Component 1 2018 Q12]*

## 2.1

# Thermochemistry

All chemical reactions involve change. These energy changes are vital to us. Plants need the energy from the Sun for the production of carbohydrates by photosynthesis: we depend on the energy content of the food we eat. The kind of life we lead depends on harnessing energy from different sources.

There are many forms of energy but basically there are only two kinds of energy, kinetic energy and potential energy. Heat is a form of kinetic energy while the energy of chemical bonds is a form of potential energy. Thermochemistry is the study of the energy changes that accompany chemical reactions.

## Topic contents

You should be able to demonstrate and apply knowledge and understanding of:

- Enthalpy change of reaction, enthalpy change of combustion and standard molar enthalpy change of formation.
- Hess's law and energy cycles.
- The concept of average bond enthalpies and how they are used to carry out simple calculations.
- How to calculate enthalpy changes.
- Simple procedures to determine enthalpy changes.

### Maths skills ≫

- Use appropriate units in calculations.
- Use an appropriate number of significant figures.
- Change the subject of an equation.
- Substitute numerical values into algebraic equations.
- Solve algebraic equations.
- Identify uncertainties in measurements.
- Plot two variables from experimental data.
- Translate information between graphical and numerical forms.

# Temperature changes

Matter possesses energy in the form of kinetic energy and potential energy.

The kinetic energy of matter is the energy of motion at a molecular level. The potential energy of matter arises from the positions of the atoms relative to one another. Bond-breaking and bond-making involve changes in potential energy.

The sum of the kinetic energy and the potential energy of all the particles in a system is the internal energy of a system.

In the course of a chemical reaction, existing bonds are broken and new ones are made. This changes the chemical energy of atoms and energy is exchanged between the chemical system and the surroundings. Frequently this leads to heat being given out to, or taken in from, the surroundings.

Most chemical reactions release energy to their surroundings. This can be detected by a rise in the temperature of the reaction mixture and the surroundings. These reactions are known as **exothermic** reactions. Examples are:

- Acids with metals
- In hand warmers (oxidation of iron)
- Thermite reaction (aluminium and iron(III) oxide).

In some reactions the system absorbs energy from its surroundings in the form of heat. Such reactions are known as **endothermic** reactions. Examples are:

- Melting ice
- In cold packs (dissolving ammonium chloride in water)
- Thermal decomposition of Group 2 carbonates.

>> **Key terms**

An **exothermic reaction** is one that releases energy to the surroundings, there is a temperature rise and $\Delta H$ is negative.

An **endothermic reaction** is one that takes in energy from the surroundings, there is a temperature drop and $\Delta H$ is positive.

**Link**

Thermal decomposition of carbonates page 70

# Enthalpy changes

The amount of heat transferred in a given chemical reaction depends on the conditions under which the reaction occurs. Most chemical reactions in the laboratory take place under constant pressure. The total energy content of a system held at constant pressure is defined as its **enthalpy**, **H**.

Just like internal energy, enthalpy cannot be measured directly. However, an **enthalpy change, $\Delta H$**, can easily be measured. Its units are joules, J, or kilojoules, kJ.

$$\Delta H = H_{products} - H_{reactants}$$

For exothermic changes (i.e a reaction that releases heat) heat is given out to the surroundings so $H_{products} < H_{reactants}$ and $\Delta H$ is negative.

For endothermic changes (i.e a reaction that absorbs heat) heat is taken in from the surroundings so $H_{products} > H_{reactants}$ and $\Delta H$ is positive.

Enthalpy changes can be represented as enthalpy profile diagrams.

>> **Key terms**

Enthalpy, **H**, is the heat content of a system at constant pressure.

Enthalpy change, $\Delta H$, is the heat added to a system at constant pressure.

>> **Study point**

The chemical system is the reactants and products.

The surroundings are everything other than the system.

Enthalpy profile diagram for an exothermic reaction ($\Delta H$ is negative):

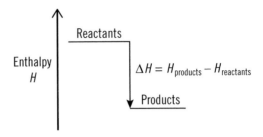

Enthalpy profile diagram for an endothermic reaction ($\Delta H$ is positive):

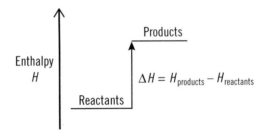

For example, when hydrogen and oxygen combine to form water:

$$H_2(g) + \tfrac{1}{2}O_2(g) \longrightarrow H_2O(g) \qquad\qquad \Delta H = -242\,kJ\,mol^{-1}$$

$$H_2(g) + \tfrac{1}{2}O_2(g) \longrightarrow H_2O(l) \qquad\qquad \Delta H = -286\,kJ\,mol^{-1}$$

This can be shown on an enthalpy diagram:

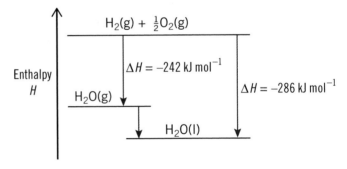

From this it can be seen that for

$$H_2O(g) \longrightarrow H_2O(l) \qquad\qquad \Delta H = -44\,kJ\,mol^{-1}$$

therefore it is an exothermic reaction.

# Conservation of energy

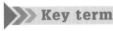
In all the examples in this chapter we have assumed that all the energy that leaves a system enters the surroundings (and vice versa). In an exothermic reaction, energy is not being created, and energy is not being destroyed in an endothermic reaction. The total energy of the whole system of reacting chemicals and surroundings is constant. This is an important principle and is called the **principle of conservation of energy**.

# Standard conditions

Since the enthalpy change for reactions depends on the conditions, for values to be compared standard enthalpy change is measured when fixed conditions are used. The conditions are:

- All substances in their standard states
- A temperature of 298 K (25 °C)
- A pressure of 1 atm (101 000 Pa).

The symbol for a standard enthalpy change is $\Delta H^\ominus$.

# Standard enthalpy change of formation, $\Delta_f H^\ominus$

This is the enthalpy change when one mole of a substance is formed from its constituent elements in their standard states under standard conditions.

For example, the standard enthalpy change of formation of carbon dioxide is represented as:

$$C(graphite) + O_2(g) \longrightarrow CO_2(g) \qquad \Delta_f H^\ominus = -394 \, kJ \, mol^{-1}$$

while the standard enthalpy change of formation of carbon monoxide is represented as:

$$C(graphite) + \tfrac{1}{2}O_2(g) \longrightarrow CO(g) \qquad \Delta_f H^\ominus = -111 \, kJ \, mol^{-1}$$

The 'per mole' refers to the formation of one mole of compound, not to the quantity of elements. Therefore the equation

$$2C(graphite) + O_2(g) \longrightarrow 2CO(g)$$

does not represent the enthalpy change of formation of carbon monoxide as two moles of carbon monoxide have formed.

If we are 'forming' an element, such as $H_2(g)$, from the element $H_2(g)$, there is no chemical change. Therefore all elements in their standard state have a standard enthalpy change of formation of $0 \, kJ \, mol^{-1}$.

> ## ›› Study point
>
> Standard state of a substance is the substance in its pure form at 1 atm and the stated temperature (normally 298 K).

> ## ›› Study point
>
> When we write thermochemical equations, we may need to use fractions like '$\tfrac{1}{2}O_2$', so that we have the correct number of moles involved in the change. In the example, the standard enthalpy change of formation refers to the carbon monoxide, so to get a balanced equation we need half a mole of oxygen molecules.

# Standard enthalpy change of combustion, $\Delta_c H^\ominus$

This is the enthalpy change when one mole of a substance is completely combusted in oxygen under standard conditions.

For example, the standard enthalpy change of combustion of methane is given by:

$$CH_4(g) + 2O_2(g) \longrightarrow CO_2(g) + 2H_2O(l) \qquad \Delta_c H^\ominus = -891 \, kJ \, mol^{-1}$$

while that of hydrogen is given by:

$$H_2(g) + \tfrac{1}{2}O_2(g) \longrightarrow H_2O(l) \qquad \Delta_c H^\ominus = -286 \, kJ \, mol^{-1}$$

This time the 'per mole' refers to the substance being combusted not to the quantity of products formed.

> ## Exam tip
>
> If an element exists in more than one state, the standard state used should be the one that is most stable at 1 atm and 298 K. For carbon, graphite is more stable than diamond so in enthalpy equations, C(graphite) should be used rather than C(s).

**1 Knowledge check**

State what is meant by the standard enthalpy change of formation, $\Delta_f H^{\ominus}$.

**2 Knowledge check**

In the reaction:

$2H_2S(g) + 3O_2(g) \longrightarrow 2SO_2(g) + 2H_2O(l)$

Explain why the standard enthalpy change of formation for $O_2(g)$ is zero.

# Enthalpy change of reaction, $\Delta_r H$

This is the enthalpy change in a reaction between the number of moles of reactants shown in the equation for the reaction.

There is no reason why a standard enthalpy change of reaction should be related to one mole of reactants. Therefore it is necessary to make clear which reaction equation is being used when an enthalpy change of reaction is being quoted.

For example, burning ethane in oxygen can be written as:

(i) $2C_2H_6(g) + 7O_2(g) \longrightarrow 4CO_2(g) + 6H_2O(l)$     $\Delta_r H = -3120\,kJ\,mol^{-1}$

(ii) $C_2H_6(g) + 3\frac{1}{2}O_2(g) \longrightarrow 2CO_2(g) + 3H_2O(l)$     $\Delta_r H = -1560\,kJ\,mol^{-1}$

Note that the value of $\Delta_r H$ in (i) is twice that of $\Delta_r H$ in (ii).

## Calculating enthalpy change of reaction

The standard enthalpy change for a chemical reaction can be calculated from the standard enthalpy changes of formation of all reactants and products involved. The standard enthalpy of reaction, $\Delta_r H^{\ominus}$, is given by:

$$\Delta_r H^{\ominus} = \Sigma \Delta_f H(\text{products}) - \Sigma \Delta_f H(\text{reactants})$$     ($\Sigma$ stands for 'sum of')

---

**Worked example**

Calculate the standard enthalpy change of reaction for:

$$CS_2(l) + 4NOCl(g) \longrightarrow CCl_4(g) + 2SO_2(g) + 2N_2(g)$$

given the following

| Compound | $CS_2(l)$ | $NOCl(g)$ | $CCl_4(g)$ | $SO_2(g)$ |
|---|---|---|---|---|
| $\Delta_f H^{\ominus}$ / kJ mol$^{-1}$ | 88 | 53 | −139 | −296 |

$\Delta_r H^{\ominus} = \Sigma \Delta_f H(\text{products}) - \Sigma \Delta_f H(\text{reactants})$

$= (-139 + 2(-296) + 2(0)) - (88 + 4(53))$     (remember that $\Delta_f H$ for any

$= -731 - 300$     element in its standard state is 0)

$= -1031\,kJ\,mol^{-1}$

---

# Hess's law

**Key term**

**Hess's law** states that the total enthalpy change for a reaction is independent of the route taken from the reactants to the products.

In the previous example, the standard enthalpy change of reaction depends only on the difference between the standard enthalpy of the reactants and the standard enthalpy of the products. However, in many chemical reactions the reactants may be able to change into the products by more than one route. Such reactions were studied by the Russian chemist Germain Hess and in 1840 he developed a chemical version of the principle of conservation of energy.

**Hess's law** states that the total enthalpy change for a reaction is independent of the route taken from the reactants to the products.

For example, consider the enthalpy cycle on the left showing two routes for converting reactants to products. The first is a direct route and the second an indirect route via the formation of an intermediate.

By Hess's law the total enthalpy is independent of the route, so route 1 = route 2

i.e.     $\Delta H_1 = \Delta H_2 + \Delta H_3$

If the sum of $\Delta H_2 + \Delta H_3$ was different from $\Delta H_1$, it would be possible to create energy by making the products via the intermediate by one route and then converting back to the reactants by the other route. This would be contrary to the law of conservation of energy.

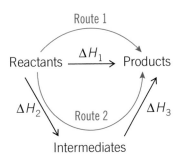

▲ Hess's law example

It follows that $\Delta H_1 = \Delta H_2 + \Delta H_3$, i.e. the standard enthalpy change is the same for the different routes. Note also that the enthalpy change of reaction from products to reactants is $-\Delta H_1$.

Hess's law enables us to calculate standard enthalpy changes of reactions that would be difficult to measure by using enthalpy cycles.

For example, forming nitrogen dioxide from nitric oxide:

$$NO(g) + \tfrac{1}{2}O_2(g) \longrightarrow NO_2(g)$$

The enthalpy changes of formation of the reactants and products are;

$$\Delta_f H\, NO = 90.3\,\text{kJ}\,\text{mol}^{-1},\ \Delta_f H\, NO_2 = 33.2\,\text{kJ}\,\text{mol}^{-1},\ \Delta_f H\, O_2 = 0\,\text{kJ}\,\text{mol}^{-1}.$$

The enthalpy cycle connecting the elements to the reactants and products is given by:

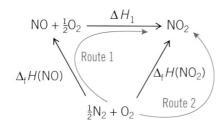

▲ Formation of NO$_2$

Since $\Delta_f H$ is given, the directions of the arrows go from the common elements to the reactants and products.

By Hess's law,  route 1 = route 2

$$\Delta H_1 + 90.3 = 33.2$$
$$\Delta H_1 = 33.2 - 90.3$$
$$\Delta H_1 = -57.1\,\text{kJ}\,\text{mol}^{-1}$$

Or use the equation

$$\Delta_r H = \Sigma\Delta_f H(\text{products}) - \Sigma\Delta_f H(\text{reactants})$$
$$\Delta_r H = 33.2 - 90.3$$
$$\Delta_r H = -57.1\,\text{kJ}\,\text{mol}^{-1}$$

To find the enthalpy change of formation of ethane

$$2C(\text{graphite}) + 3H_2(g) \longrightarrow C_2H_6(g)$$

is impossible practically since burning carbon in hydrogen will produce a mixture of hydrocarbons. However, the enthalpy change of combustion of carbon, $\Delta_c H(C) = -394\,\text{kJ}\,\text{mol}^{-1}$, hydrogen, $\Delta_c H(H_2) = -286\,\text{kJ}\,\text{mol}^{-1}$, and ethane, $\Delta_c H(C_2H_6) = -1560\,\text{kJ}\,\text{mol}^{-1}$, can be measured accurately. Drawing an enthalpy cycle and Hess's law can be used to calculate the enthalpy change of formation of ethane.

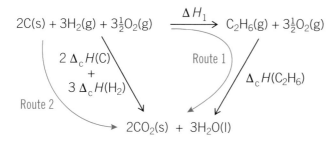

▲ Formation of ethane

**4** **Knowledge check**

Given the following enthalpy changes of combustion

| Substance | $C_2H_2(g)$ | $H_2(g)$ | $C_2H_6(g)$ |
|---|---|---|---|
| $\Delta_cH^{\ominus}/$ kJ mol$^{-1}$ | −1301 | −286 | −1560 |

Calculate the enthalpy change for the reaction

$C_2H_2(g) + 2H_2(g) \longrightarrow C_2H_6(g)$

Since $\Delta_cH$ is given, the directions of the arrows go from the reactants and products to the common combustion products.

By Hess's law,     route 1 = route 2

$$\Delta H_1 + (-1560) = (2(-394) + 3(-286))$$

$$\Delta H_1 - 1560 = -1646$$

$$\Delta H_1 = -1646 + 1560$$

$$\Delta H_1 = -86 \text{ kJ mol}^{-1}$$

Or use the equation:     $\Delta_rH = \Sigma\Delta_cH\text{(reactants)} - \Sigma\Delta_cH\text{(products)}$

$$\Delta H = (2(-394) + 3(-286)) - (-1560)$$

$$\Delta H = -1646 + 1560$$

$$\Delta H = -86 \text{ kJ mol}^{-1}$$

# Bond enthalpies

A way of calculating the enthalpy changes of reactions involving covalent compounds is to consider the enthalpy associated with each covalent bond. The amount of energy needed to break a covalent bond is called the **bond enthalpy**. The values of bond enthalpies are always positive because breaking a bond requires energy.

For example, the bond enthalpy of $HCl(g)$ is 432 kJ mol$^{-1}$ and is the enthalpy change for the process, $HCl(g) \longrightarrow H(g) + Cl(g)$.

The actual value of the bond enthalpy for a particular bond depends on the structure of the rest of the molecule, so a C–C bond in ethane, $C_2H_6$, has a slightly different value to a C–C bond in pentane, $C_5H_{12}$. Although the bond enthalpy of a given bond is similar in a wide range of compounds, **average bond enthalpies** calculated using the values from many different compounds are used.

Calculations of enthalpy changes of reactions based on average bond enthalpy values will not be as accurate as results derived from experiments with specific molecules. However, they usually give an accurate enough indication of the standard enthalpy change of reaction.

**Key terms**

**Bond enthalpy** is the enthalpy required to break a covalent X — Y bond into X atoms and Y atoms, all in the gas phase.

**Average bond enthalpy** is the average value of the enthalpy required to break a given type of covalent bond in the molecules of a gaseous species.

**Study point**

Breaking bonds requires energy, therefore is endothermic (and bond enthalpy is always positive).

Making bonds releases energy, so is exothermic and bond enthalpy is always negative.

## Calculations using bond enthalpies

There are four steps involved:

**Step 1**  Draw out each molecule to show the bonds (if not already given).

**Step 2**  Calculate the energy required to break all the bonds in the reactants (endothermic).

**Step 3**  Calculate the energy released in forming all the bonds in the products (exothermic).

**Step 4**  Use the equation

Enthalpy change of reaction = Total energy needed to break bonds − total energy released in forming bonds

$\Delta H = \Sigma \text{ (bonds broken)} - \Sigma \text{ (bonds formed)}$

## Worked example

Using average bond enthalpies calculate the standard enthalpy change of reaction for the complete combustion of methane

$$CH_4(g) + 2O_2(g) \longrightarrow CO_2(g) + 2H_2O(g)$$

| Bond | C–H | O=O | C=O | O–H |
|------|-----|-----|-----|-----|
| Average bond enthalpy/ kJ mol$^{-1}$ | 413 | 496 | 805 | 463 |

**Step 1**  Draw out each molecule

**Step 2**  Calculate the energy required to break the bonds (endothermic).

Bonds broken:

$$4(C-H) + 2(O=O) = (4 \times 413) + (2 \times 496) = 2644 \, kJ \, mol^{-1}$$

**Step 3**  Calculate the energy released when bonds are made (exothermic).

Bonds formed:

$$2(C=O) + 4(O-H) = (2 \times 805) + (4 \times 463) = 3462 \, kJ \, mol^{-1}$$

**Step 4**  Use the equation:

$$\Delta H = \Sigma \text{(bonds broken)} - \Sigma \text{(bonds formed)}$$

$$\Delta H = 2644 - 3462 = -818 \, kJ \, mol^{-1}$$

### ≪ Exam tip

When asked to calculate enthalpy change using bond enthalpies, always draw out each molecule so that you can see the bonds broken and bonds made.

### Knowledge check 5

Using average bond enthalpies, calculate the enthalpy change for the reaction

$$2H_2(g) + O_2(g) \longrightarrow 2H_2O(g)$$

| Bond | H–H | O=O | O–H |
|------|-----|-----|-----|
| Average bond enthalpy/ kJ mol$^{-1}$ | 436 | 496 | 463 |

### Knowledge check 6

Chlorine can react with ethane to give chloroethane

$$C_2H_6 + Cl_2 \longrightarrow C_2H_5Cl + HCl$$

$\Delta H = -123 \, kJ \, mol^{-1}$

Calculate the average bond enthalpy, in kJ mol$^{-1}$, for the Cl–Cl bond.

| Bond | C–C | C–H | C–Cl | H–Cl |
|------|-----|-----|------|------|
| Average bond enthalpy/ kJ mol$^{-1}$ | 348 | 413 | 346 | 432 |

# Measuring enthalpy changes

You cannot directly measure the heat content (enthalpy) of a system but you can measure the heat transferred to its surroundings. This process involves carrying out the chemical change in an insulated container called a calorimeter. The change in the temperature inside the calorimeter caused by the enthalpy change of the reaction can be measured with a thermometer.

If the temperature change is recorded, and the mass and specific heat capacity of the contents of the calorimeter are known, then the enthalpy change can be calculated.

The specific heat capacity is the heat required to raise the temperature of 1 g of a substance by 1 K. The value for water is $4.18 \, J \, g^{-1} \, K^{-1}$.

The relationship between the temperature change, $\Delta T$, and the amount of heat transferred, $q$, is given by the expression:

$$q = mc\Delta T$$

where $m$ is the mass of the solution

and $c$ is the specific heat capacity of the solution.

For the purposes of the calculations, we assume that all the heat is exchanged with the solution alone, the solution has the same specific heat capacity as water and the density of the solution is $1 \, g \, cm^{-3}$.

### ≫ Study point

A calorimeter is any vessel used for determinations of heat changes. It is named after the old unit of heat, the calorie (1 cal = 4.18 J).

### ≫ Study point

In the expression $q = mc\Delta T$, you do not need to change °C into K, because temperature change is being used not actual temperatures.

The minus sign is used in the expression $\Delta H = -q/n$ because if there is an increase in temperature, the reaction is exothermic and $\Delta H$ is negative.

Therefore $q = m \times 4.18 \times \Delta T$ and the mass will be the same as the volume of the solution. To obtain the maximum temperature change, allowances are made for heat lost (or gained) to the surroundings. Therefore, temperatures of the solution are taken for a short period before mixing and for some time after mixing. A graph of temperature against time is plotted and the maximum temperature is obtained by extrapolating the graph back to the mixing time.

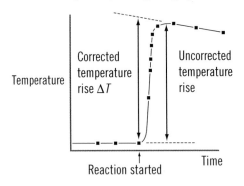

▲ Graph of temperature vs time

To calculate the enthalpy change of reaction per mole, we use the expression:

$$\Delta H = \frac{-q}{n}$$

where $n$ is the amount in moles that has reacted.

**Worked examples**

**1 Displacement reactions**

$6\,g$ of zinc were added to $25.0\,cm^3$ of $1.00\,mol\,dm^{-3}$ copper(II) sulfate solution in a polystyrene cup. The temperature increased from $20.2\,°C$ to $70.8\,°C$.

Calculate the enthalpy change for the reaction

$$Zn(s) + CuSO_4(aq) \longrightarrow ZnSO_4(aq) + Cu(s)$$

(Assume that the density of the solution is $1.00\,g\,cm^{-3}$ and its specific heat capacity, $c$, is $4.18\,J\,g^{-1}\,K^{-1}$.)

**Step 1**   Calculate the amount of heat transferred in the experiment.

$25\,cm^3$ of solution, therefore mass is $25\,g$

$$\Delta T = 70.8 - 20.2$$

$$q = mc\Delta T = 25 \times 4.18 \times 50.6 = 5288\,J$$

**Step 2**   Calculate the amount, in moles, of the reactants

$$\text{moles Zn} = \frac{m}{M} = \frac{6}{65.4} = 0.092$$

$$\text{moles CuSO}_4 = c \times \frac{v}{1000} = 1 \times 0.025 = 0.025$$

therefore $CuSO_4$ is not in excess and is the amount, $n$, used in the calculation.

**Step 3**   Calculate the molar enthalpy change

$$\Delta H = \frac{-q}{n} = \frac{-5288}{0.025} = -211\,520\,J\,mol^{-1}$$

$$= -211.5\,kJ\,mol^{-1}$$

**2 Neutralisation reactions**

$25.0\,cm^3$ of HCl(aq) are added to $25.0\,cm^3$ of NaOH(aq), both of concentration $1.00\,mol\,dm^{-3}$, in an insulated container. The acid and alkali are completely neutralised and the maximum temperature rise was calculated as $6.4\,°C$.

Calculate the molar enthalpy change of neutralisation for the reaction

$$HCl(aq) + NaOH(aq) \longrightarrow NaCl(aq) + H_2O(l)$$

(Assume that the density of the solution is $1.00 \, gcm^{-3}$ and its specific heat capacity, $c$, is $4.18 \, Jg^{-1}K^{-1}$.)

**Step 1**    Calculate the amount of heat transferred in the experiment.

Since $25.0 \, cm^3$ of acid and alkali used, total volume of solution is $50.0 \, cm^3$

$$q = mc\Delta T = 50 \times 4.18 \times 6.4 = 1338 \, J$$

**Step 2**    Calculate the amount, in moles, of the reactants:

$$n = cv = 1.0 \times \frac{25.0}{1000} = 0.025$$

**Step 3**    Calculate the molar enthalpy change:

$$\Delta H = \frac{-q}{n} = \frac{-1338}{0.025} = -53520 \, Jmol^{-1}$$
$$= -53.5 \, kJmol^{-1}$$

**《 Maths tip**

Don't forget to divide by 1000 to convert 'J' to 'kJ'.

## 3 Dissolving a solid to form an aqueous solution

$7.80 \, g$ of ammonium nitrate, $NH_4NO_3$, were dissolved in $50.0 \, cm^3$ of water in an insulated container. The temperature fell by $11.1 \, °C$. Calculate the enthalpy of solution for the process:

$$NH_4NO_3(s) + aq \longrightarrow NH_4^+(aq) + NO_3^-(aq)$$

(Assume that the density of the solution is $1.00 \, gcm^{-3}$ and its specific heat capacity, $c$, is $4.18 \, Jg^{-1}K^{-1}$.)

**Step 1** Calculate the amount of heat transferred in the experiment:

$$q = mc\Delta T = 50 \times 4.18 \times (-11.1) = -2320 \, J$$

**Step 2** Calculate the amount, in moles, of the reactant:

$$n = \frac{m}{M} = \frac{7.80}{80.04} = 0.0975$$

**Step 3** Calculate the molar enthalpy change:

$$\Delta H = \frac{-q}{n} = \frac{-(-2320)}{0.0975} = 23795 \, Jmol^{-1}$$
$$= 23.8 \, kJmol^{-1}$$

## 4 Combustion of alcohol

The combustion of $0.750 \, g$ of ethanol raised the temperature of $250 \, cm^3$ of water by $19.5 \, °C$. Calculate the enthalpy change of combustion of ethanol, $C_2H_5OH$.

(The specific heat capacity, $c$, of water is $4.18 \, Jg^{-1}K^{-1}$.)

**Step 1** Calculate the amount of heat transferred in the experiment:

$$q = mc\Delta T = 250 \times 4.18 \times 19.5 = 20378 \, J$$

**Step 2** Calculate the amount, in moles, of the reactant:

$$n = \frac{m}{M} = \frac{0.75}{46.06} = 0.0163$$

**Step 3** Calculate the molar enthalpy change:

$$\Delta H = \frac{-q}{n} = \frac{-20378}{0.0163} = -1250184 \, Jmol^{-1}$$
$$= -1250 \, kJmol^{-1}$$

**Knowledge check** 8 ◀

40.0 $cm^3$ of HCl(aq) are added to 40.0 $cm^3$ of NaOH(aq), both of concentration $0.80 \, moldm^{-3}$, in an insulated container. The acid and alkali are completely neutralised and the maximum temperature rise was calculated at $5.2 \, °C$.

Calculate the molar enthalpy change of neutralisation for the reaction.

(The specific heat capacity, $c$, of water is $4.18 \, Jg^{-1}K^{-1}$.)

# Practical activity

## Combustion

Measurements of enthalpy changes of combustion are important as they help to compare the energy available from the oxidation of different flammable liquids that may be used as fuels.

The diagram below shows a simple method for obtaining an approximate value for the enthalpy of combustion of a fuel such as an alcohol.

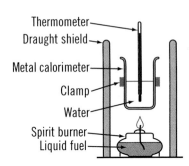

▲ Alcohol combustion

Since the mass of the water is used in the expression to calculate $\Delta H$, it must be measured accurately. The number of moles is needed to calculate the molar enthalpy change, therefore the mass of fuel burned also has to be measured accurately.

These are the main points to note in the practical work:

- Allow a suitable gap between the base of the metal container and the top of the spirit burner.
- Accurately measure the amount of water being added to the metal container.
- Use an accurate thermometer to measure the initial temperature of the water. When a steady value has been obtained, record the temperature.
- Weigh the spirit burner containing the alcohol and record the initial mass.
- After lighting the wick, adjust the gap between the metal container and the spirit burner if necessary.
- Allow the alcohol to heat the water to a suitable temperature (an increase of about 20 °C is adequate – the smaller the increase, the greater the error on the thermometer).
- Extinguish the flame and record the final maximum temperature.
- Allow the spirit burner to cool thoroughly before reweighing and recording the final mass.

The $\Delta H$ value obtained will be much lower than the book value because:

- Some of the energy transferred from the burning alcohol is 'lost' in heating the apparatus and the surroundings.
- The alcohol is not completely combusted.
  (Soot (carbon) on the bottom of the calorimeter will show whether this has happened.)

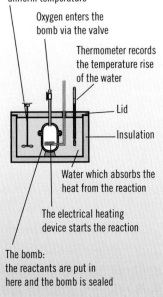
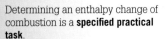

# Indirect determination of an enthalpy change

The simplest type of calorimeter is a coffee cup calorimeter. It can be used to measure changes that take place in aqueous solution.

The expanded polystyrene insulates the solution inside the cup so the amount of heat lost or absorbed by the cup during the experiment is negligible. (The cup can be placed inside another polystyrene cup or inside a beaker and lagged with cotton wool to improve insulation.)

Since the enthalpy change of many reactions, especially those involving solids (and gases), can be difficult to measure directly, solids are reacted with acids to form solutions. A coffee cup calorimeter can then be used.

All actual values obtained this way are lower than book values due to heat loss from the simple type of calorimeter used.

An example of indirect determination of enthalpy changes is the enthalpy change of reaction of magnesium oxide with carbon dioxide.

The enthalpy changes of reaction between the solid reactant (magnesium oxide) and acid and the solid product (magnesium carbonate) and acid are separately measured (hydrochloric acid is a suitable acid). The heat evolved is calculated, corrected for heat loss to the surroundings, scaled up to molar amounts of the solids involved and Hess's law is used to obtain the desired enthalpy change.

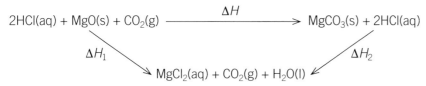

$2HCl(aq) + MgO(s) + CO_2(g) \xrightarrow{\Delta H} MgCO_3(s) + 2HCl(aq)$

$\Delta H_1 \searrow \qquad \swarrow \Delta H_2$

$MgCl_2(aq) + CO_2(g) + H_2O(l)$

These are the main points to note in the practical work:

- Measure an appropriate volume of acid using a burette or pipette (the mass of acid is used in the expression to calculate $\Delta H$) and place the acid in a polystyrene cup. The acid must be in excess (to ensure that all the solid reacts).

- Use an accurate thermometer to measure the initial temperature of the acid. When a steady value has been obtained, record the temperature. ($\Delta T$ is used in the expression to calculate $\Delta H$.)

- Accurately weigh the solid, in powder form (to ensure as rapid a reaction as possible), in a suitable container (the amount, in moles, of the solid is used in the expression to calculate $\Delta H$).

- Add all the solid to the cup, stir the mixture well (to ensure that the reaction is as rapid as possible and that all the solid is used) and start a stop-watch.

- Keep stirring with the thermometer and record the temperature regularly (about every 30 seconds). Stop recording the temperature when it has fallen for about 5 minutes.

- Re-weigh the weighing container to ensure that the correct mass of solid added is recorded.

- Plot a graph of temperature against time to calculate the maximum temperature the mixture might have reached. (This is essential to calculate the correct $\Delta T$ – see page 104.)

- Calculate the amount of heat transferred ($q = mc\Delta T$).

- Calculate the enthalpy change for the reaction ($\Delta H = -q/n$).

- Repeat the procedure with the other solid.

- Use Hess's law to calculate the required enthalpy change ($\Delta H = \Delta H_1 - \Delta H_2$).

This process is also suitable to find the enthalpy change of reaction for displacement reactions and for dissolving solids in water to find the enthalpy change of solution.

▲ Coffee cup calorimeter

**Knowledge check** 9

When using a coffee cup calorimeter to determine the enthalpy change of a displacement reaction, explain why:

(a) The solid used is in powdered form.

(b) A graph of temperature against time is plotted and the graph extrapolated back to the mixing time.

**◀◀ Practical check**

Indirect determination of an enthalpy change of reaction is a **specified practical task**.

Exact details of indirect determinations will differ, but the procedures will remain the same.

Other suitable examples are:

Formation of magnesium oxide (adding magnesium and magnesium oxide separately to hydrochloric acid).

Formation of magnesium carbonate (adding magnesium and magnesium carbonate separately to hydrochloric acid).

Decomposition of sodium hydrogencarbonate (adding sodium hydrogencarbonate and sodium carbonate separately to hydrochloric acid).

# Test yourself

1. (a) State what is meant by the term standard molar enthalpy change of combustion. [2]

   (b) Write the equation for the complete combustion of octane, $C_8H_{18}$. [1]

   (c) For the energy cycle

$$4C(s) + 5H_2(g) + 6\tfrac{1}{2}O_2(g) \xrightarrow{\;\;\Delta H\;\;} C_4H_{10}(g) + 6\tfrac{1}{2}O_2(g)$$

$$4CO_2(g) + 5H_2O(l)$$

   Use the values in the table below to calculate the enthalpy change of reaction, $\Delta H$. [2]

| Substance | Enthalpy change of combustion, $\Delta_c H^\ominus$ / kJ mol$^{-1}$ |
|---|---|
| Carbon | −394 |
| Hydrogen | −286 |
| Butane | −2878 |

   (d) (i) Write an equation to represent the standard molar enthalpy change of formation, $\Delta_f H^\ominus$, of $Cl_2O(g)$ [1]

   (ii) The standard molar enthalpy change of formation, $\Delta_f H^\ominus$, of $Cl_2O(g)$ is 76.2 kJ mol$^{-1}$. Using this value and the average bond enthalpies given in the table below, calculate the average bond enthalpy of the Cl–O bond in $Cl_2O$. [2]

| Bond | Average bond enthalpy / kJ mol$^{-1}$ |
|---|---|
| Cl–Cl | 242 |
| O=O | 496 |

2. 50.0 cm$^3$ of aqueous potassium hydroxide solution of concentration 0.500 mol dm$^{-3}$ and 25.0 cm$^3$ of aqueous sulfuric acid solution of concentration 0.500 mol dm$^{-3}$ were mixed in a polystyrene cup. The initial temperature of the solutions was 18.5°C and the temperature rose to 23.0°C.

   Calculate the molar enthalpy change of neutralisation for the reaction

$$H_2SO_4 + 2KOH \longrightarrow K_2SO_4 + 2H_2O$$ [5]

3. (a) State Hess's Law. [1]

   (b) Glucose burns in oxygen to produce carbon dioxide and water. The equation for the reaction is

$$C_6H_{12}O_6(s) + 6O_2(g) \longrightarrow 6CO_2(g) + 6H_2O(l)$$

   (i) Use the standard enthalpy change of formation values, $\Delta_f H^\ominus$, given in the table below to calculate the standard enthalpy change, $\Delta_f H^\ominus$, for this reaction. [2]

| Species | $\Delta_f H^\ominus$ / kJ mol$^{-1}$ |
|---|---|
| $C_6H_{12}O_6(s)$ | −1271 |
| $O_2(g)$ | 0 |
| $CO_2(g)$ | −394 |
| $H_2O(l)$ | −286 |

   (ii) Explain why the standard enthalpy change of formation for $O_2(g)$ is zero. [1]

4. (a) Edmund used the apparatus below to find the enthalpy change of combustion of heptane, $C_7H_{16}$.

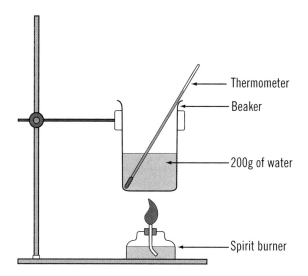

Thermometer
Beaker
200g of water
Spirit burner

(i) During the experiment the temperature of the water rose by 28.2°C.

He calculated the value to be –3280 kJ mol⁻¹.

Calculate the mass of heptane that he used in this experiment. [5]

(ii) Give one reason why the experimental value that Edmund obtained differs greatly from the literature value. Suggest an improvement to the experiment that would give a more accurate value. [2]

(b) Use the following table to estimate the actual enthalpy change of combustion of heptane. Explain fully how you obtained your answer. [2]

| Alkane | Enthalpy change of combustion / kJ mol⁻¹ |
|---|---|
| pentane, $C_5H_{12}$ | –3509 |
| nonane, $C_9H_{20}$ | –6125 |
| decane, $C_{10}H_{22}$ | –6779 |

(c) Use the values in the table above to show that pentane gives more energy per gram of fuel burned than decane. [2]

(d) (i) Another way of calculating enthalpy change of reactions is by using average bond enthalpies.

Use the values in the table below to calculate the enthalpy change for the combustion of heptane. [4]

$$C_7H_{16}(l) + 11O_2(g) \longrightarrow 7CO_2(g) + 8H_2O(g)$$

| Bond | Average bond enthalpy / kJ mol⁻¹ |
|---|---|
| C–C | 348 |
| C–H | 412 |
| O=O | 496 |
| C=O | 743 |
| O–H | 463 |

(ii) Give a reason why the calculated value in (d)(i) is different from the actual value obtained in (b). [1]

# Rates of reaction

Many people are interested in knowing how to alter the rates of chemical reactions. Fertiliser manufacturers want to speed up the formation of ammonia. Car manufacturers want to slow down the rate at which iron rusts. Chemists seeking to manage environmental issues such as ozone depletion need to understand reaction rates.

This topic is concerned with measuring reaction rates in the laboratory, recognising what factors affect reaction rates and understanding how these factors affect reaction rates.

## Topic contents

You should be able to demonstrate and apply knowledge and understanding of:

- How to calculate rates from experimental data and how to establish the relationship between reactant concentrations and rate.
- Collision theory in explaining the effects of changing conditions on reaction rate.
- The concepts of energy profiles and activation energy.
- The rapid increase in rate with temperature in terms of changes in the Boltzmann energy distribution curve.
- The characteristics of a catalyst.
- How catalysts increase reaction rates by providing an alternative route of lower activation energy.
- How colorimetry can be used in studies of some reaction rates.
- Measurements of reaction rate by gas collection and precipitation methods and by an 'iodine clock' reaction.

### Maths skills »

- Recognise and use numbers in decimal, ordinary and standard form.
- Use appropriate units in calculations.
- Use an appropriate number of significant figures.
- Identify uncertainties in measurements.
- Plot two variables from experimental data.
- Translate information between graphical and numerical forms.

For chemical reactions, the **rate of a reaction** is found by measuring the amount of a reactant used up (or product formed) per unit of time. The amount is often expressed in terms of concentration.

For a reaction: rate $= \dfrac{\text{change in concentration}}{\text{time}}$ units: $\dfrac{\text{mol dm}^{-3}}{\text{s}} = \text{mol dm}^{-3}\,\text{s}^{-1}$

If another variable, such as mass or volume is measured, the rate can be expressed in corresponding units such as $\text{g s}^{-1}$ or $\text{cm}^3\,\text{s}^{-1}$.

However, it is obvious from watching a reaction that the instantaneous rate changes as the reaction proceeds.

Usually for reactions:

- Rate is fastest at the start of a reaction since each reactant has its greatest concentration.
- Rate slows down as the reaction proceeds since the concentration of the reactants decreases.
- Rate becomes zero when the reaction stops, i.e. when one of the reactants has been used up.

>>> **Key term**

The **rate of reaction** is the change in concentration of a reactant or product per unit time.

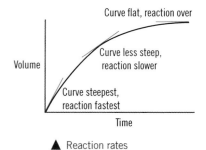

▲ Reaction rates

# Calculating rates

We follow the rate of a reaction by measuring the amount of a reactant used up or product formed over a period of time. The results obtained are plotted to give a graph. To find the initial rate, it is necessary to find the gradient (initial slope) of the line. If the graph is a curve, you have to draw a tangent to the curve and find the gradient of the tangent.

Gradient $= \dfrac{\text{change in the vertical height}}{\text{change in the horizontal line}} = \dfrac{\Delta y}{\Delta x}$

**Maths tip**

A tangent is a straight line that touches the curve and has the same gradient as the curve at that point.

## Worked examples

### Example 1

Find the initial rate of reaction for this straight line graph.

**Answer**

To find the initial rate at any convenient point, P, on the straight line draw a horizontal line MP to the $y$ axis and draw a vertical line from M to the beginning of the slope, N.

Initial rate = gradient =

$\dfrac{\Delta y}{\Delta x} = \dfrac{\text{N} - \text{M}}{\text{P} - \text{M}} = \dfrac{(0.98 - 0.79)}{22 - 0} =$

$\dfrac{0.19}{22} = 0.0086\ \text{mol dm}^{-3}\ \text{s}^{-1}$

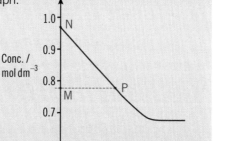

### Example 2

Find the initial rate of reaction for this curved line graph.

**Answer**

Since the graph is a curve, to find the initial rate a tangent must be drawn at $t = 0$ and the gradient of the tangent calculated.

Initial rate = gradient =

$\dfrac{\Delta y}{\Delta x} = \dfrac{\text{M} - \text{N}}{\text{P} - \text{M}} = \dfrac{0.26}{10} = 0.026\ \text{mol dm}^{-3}\ \text{s}^{-1}$

To find the rate of a reaction at a specific time, a tangent is drawn at that time e.g. to find the rate after 10 seconds, draw a tangent at $t = 10\text{s}$ and calculate the gradient.

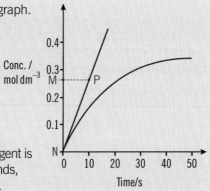

**Knowledge check** 2.1 ◀

The volume-time graph for a reaction is shown below:

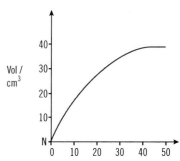

Calculate the rate of reaction after 20 seconds. State the unit.

To find out the relationship between initial rate and the initial concentrations of the reactants, a series of experiments, in which the concentration of only one reactant is changed at a time, must be performed.

$$\text{Since rate} = \frac{\text{change in concentration}}{\text{time}} \qquad \text{rate} \propto \frac{1}{\text{time}}$$

If a graph of $1/t$ is plotted against concentration, the relationship between reactant concentration and rate can be established.

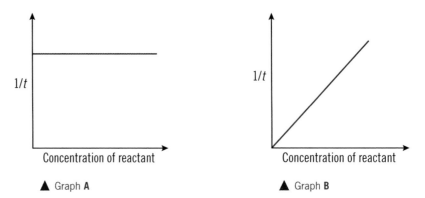

▲ Graph **A**    ▲ Graph **B**

In graph **A** although the concentration of the reactant has changed, the rate does not change, therefore the rate of reaction is independent of the concentration of that particular reactant.

In graph **B** as the concentration of the reactant increases, the rate also increases, therefore the rate of reaction is directly proportional to the concentration of that particular reactant.

If the results are tabulated, the results of the initial concentrations of reactants and rates can be compared.

For example, the table below gives the experimental data for the reaction between propanone, $CH_3COCH_3$, and iodine, $I_2$, carried out in dilute hydrochloric acid.

$$CH_3COCH_3(aq) + I_2(aq) \longrightarrow CH_3COCH_2I(aq) + HI(aq)$$

| Experiment | Initial concentrations / mol dm$^{-3}$ | | | Initial rate / $10^{-4}$ mol dm$^{-3}$ s$^{-1}$ |
|---|---|---|---|---|
| | $I_2$(aq) | $CH_3COCH_3$(aq) | HCl(aq) | |
| 1 | 0.0005 | 0.4 | 1.0 | 0.6 |
| 2 | 0.0010 | 0.4 | 1.0 | 0.6 |
| 3 | 0.0010 | 0.8 | 1.0 | 1.2 |

In experiments 1 and 2, only the concentration of iodine is changed and when it is doubled there is no change in the initial rate of reaction. Therefore the initial rate of reaction is independent of the initial concentration of aqueous iodine.

In experiments 2 and 3, only the concentration of propanone is changed and when it is doubled the initial rate of reaction also doubles. Therefore the initial rate of reaction is directly proportional to the initial concentration of aqueous propanone.

---

## Study point

You already know that

$$\text{speed} = \frac{\text{distance travelled}}{\text{time}}$$

So a car that travels 100 miles in two hours has an overall average speed of

$$\frac{100 \text{ miles}}{2 \text{ hours}} = 50 \text{ mph}.$$

In a chemical reaction in which the concentration of a reactant changes from $1.20 \text{ mol dm}^{-3}$ to $1.00 \text{ mol dm}^{-3}$ in 30 seconds, the overall reaction rate is:

$$
\begin{aligned}
\text{Rate} &= \frac{\text{change in concentration}}{\text{time}} \\
&= \frac{(1.20 - 1.00) \text{ mol dm}^{-3}}{30 \text{ s}} \\
&= 6.7 \times 10^{-3} \text{ mol dm}^{-3} \text{ s}^{-1}
\end{aligned}
$$

---

**2** **Knowledge check**

Sodium thiosulfate reduces iodine to iodide ions. The rate of this reaction depends on the concentration of thiosulfate ions. At the beginning of a reaction between sodium thiosulfate and iodine, the concentration of thiosulfate ions was $0.100 \text{ mol dm}^{-3}$. After three seconds the concentration had fallen to $0.088 \text{ mol dm}^{-3}$. Calculate the initial rate of reaction.

# Collision theory

The explanation of rates of reaction is based on collision theory.

Collision theory says that for a reaction between two molecules to occur, an effective collision must take place, i.e. a collision that results in the formation of product molecules. The reaction rate is a measure of how frequently effective collisions occur.

Not all collisions between molecules result in reactions; however, the greater the number of collisions, the higher the chance that some of the collisions will be effective. For a collision to be effective, the molecules must collide in the correct orientation and with enough energy to react. Any factor that increases the rate of effective collisions will also increase the rate of the reaction.

# Factors that affect the rate of reaction

We know that a variety of factors can affect the position of equilibrium; similarly several factors may affect the rate of a chemical reaction.

### 1 Concentration (pressure for gases)

Increasing the concentration of reactants increases the rate of reaction, e.g. adding magnesium to concentrated acid will produce a more vigorous effervescence of hydrogen than will adding magnesium to dilute acid.

For reactions involving gases, increasing pressure will increase the rate of reaction, as pressure is proportional to concentration.

If there is an increase in concentration of reactants, there are more molecules in a given volume. Distances between the molecules are reduced so there is an increase in the number of collisions per unit time. This means that there is a greater chance that the number of effective collisions increases, hence the rate of reaction increases.

For a gaseous reaction, increasing the pressure is the same as increasing the concentration.

### 2 Temperature

Increasing the temperature of reactants increases the rate of reaction, e.g. milk turns sour more quickly on a hot summer's day than in the cold winter.

If there is an increase in temperature there will be an increase in kinetic energy of the molecules and so they will move faster. This means that more molecules will have enough energy to react on collision, meaning that the reaction rate will increase.

### 3 Particle size

In a reaction involving a solid, breaking down the solid into smaller pieces increases the rate of reaction, e.g. powdered magnesium produces hydrogen far more quickly than magnesium ribbon when added to acid.

Reducing the particle size of a solid increases the surface area so again the molecules are closer together and there is an increase in the number of collisions per unit time leading to an increase in reaction rate.

### 4 Catalysts

A catalyst increases the rate of a chemical reaction without itself undergoing a permanent change, e.g. manganese(IV) oxide helps hydrogen peroxide decompose at room temperature.

### 5 Light

Some reactions are much more vigorous when carried out in bright light, e.g. photochlorination of methane.

**Study point**

In collision theory, the term 'molecules' is used for simplicity. The species that collide could be molecules, atoms or ions.

**Exam tip**

When explaining the effect of changing conditions on reaction rates, always use collision theory in your answer. Bullet points can be useful.

**Knowledge check** 3

Explain why magnesium reacts faster when it is added to 2.0 mol dm$^{-3}$ hydrochloric acid than to 1.0 mol dm$^{-3}$ hydrochloric acid.

**Link**

Position of equilibrium pages 78–80

**Link**

Photochlorination of methane page 147

# Energy profiles and distribution curves

## Activation energy

Collision theory assumes that for a reaction to occur molecules must collide in a favourable orientation and they must collide with enough energy. This minimum kinetic energy is known as the **activation energy, $E_a$**, and it varies from one reaction to another. If the reactants collide with an energy at least equal to the activation energy, the collision is successful and products will form.

The activation may be shown on diagrams called energy profiles. These diagrams compare the enthalpy of the reactants with the enthalpy of the products:

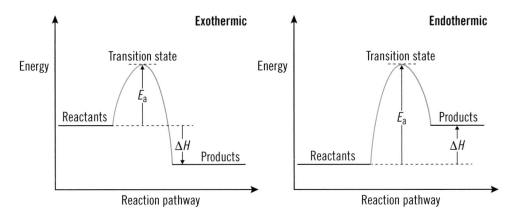

▲ Energy profile diagrams

These diagrams illustrate the role of activation energy as an energy barrier that must be overcome by reactants before they can form products.

In the first part of the energy profiles (coloured blue in the diagrams), the reactant molecules are coming together and breaking apart. Separating atoms in the reactant molecules requires bonds to be broken, so energy is absorbed.

In the second part of the energy profile (coloured orange in the diagrams), the product molecules are forming and moving apart. Producing product molecules involves forming bonds, so energy is released.

Therefore, even if a reaction is exothermic an input of energy is required to break bonds to start the reaction. Once the reaction has started, enough heat energy is produced to keep the reaction going.

The difference between the energy of the reactants and products is the enthalpy change of reaction. As seen from the profile, for an exothermic reaction the products have a lower energy than the reactants and so heat is given out. For an endothermic reaction the products have a higher energy than the reactants so heat is taken in from the surroundings.

The energy profile below shows the implication of this for a reversible reaction:

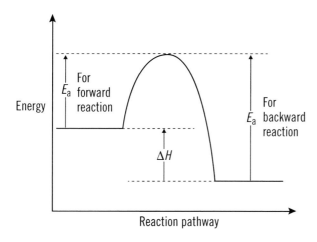

▲ Energy profile for a reversible reaction

If the reversible reaction is exothermic in one direction, it will be endothermic in the reverse direction. The enthalpy change of reaction is given by:

$$\Delta H = E_a f - E_a b$$

where $E_a f$ and $E_a b$ are the activation energies of the forward and backward reactions respectively.

For an exothermic reaction $E_a f < E_a b$ and $\Delta H$ is negative.

For an endothermic reaction $E_a f > E_a b$ and $\Delta H$ is positive.

**‹ Link ›**

Enthalpy changes page 97

**Knowledge check** 4

Calculate the enthalpy change for the reaction

$$2HI(g) \rightleftharpoons H_2(g) + I_2(g)$$

given that the activation energy for the forward reaction is $195\,kJ\,mol^{-1}$ and the activation energy for the backward reaction is $247\,kJ\,mol^{-1}$.

# The Boltzmann distribution

The molecules in a gas are moving constantly at different speeds so the energy of each molecule varies greatly. Since the molecules collide frequently, a molecule that has been knocked can move quicker with greater energy than before while the molecule that caused the collision will slow down and have hardly any energy.

The Boltzmann energy distribution curve shows the distribution of molecular energies in a gas. The energies of a few molecules are almost zero. Most molecules have an energy around an average value. Only a minority have values that equal or exceed the activation energy.

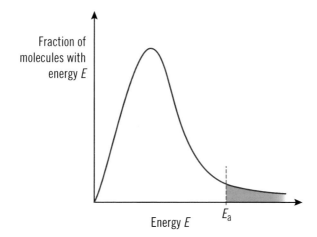

▲ Diagram of Boltzmann distribution

**≪ Exam tip**

When explaining the effect of temperature on reaction rate you must include a reference to activation energy in your answer.

At higher temperatures the average molecular energy will increase. Some molecules will still be almost motionless but at any one time many more molecules will have a higher energy. At two different temperatures, $T_1$ and $T_2$, where $T_2 > T_1$, the Boltzmann distribution curve shows why reaction rate is so dependent on temperature.

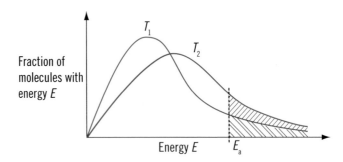

▲ Boltzmann distribution for two temperatures

- Both curves start at the origin but do not touch the energy axis again.
- The areas under the two curves are equal and are proportional to the total number of molecules in the sample.
- At the higher temperature, $T_2$, the distribution flattens, the peak moves to the right (higher energy) with a lower height.
- At the higher temperature, $T_2$, the mean energy of the molecules increases. There is a wider spread of values.
- Only the molecules with an energy equal to or greater than the activation energy, $E_a$, are able to react.
- At the higher temperature, $T_2$, many more molecules have sufficient energy to react and so the rate increases significantly.

**Knowledge check**

In the reaction between magnesium and hydrochloric acid, explain why raising the temperature of the acid increases the rate of reaction.

# Catalysts

Many reactions have high activation energies. This means that in order to convert reactants to products at a reasonable rate, the reaction mixture needs to be maintained at a very high temperature. However, this difficulty may be overcome by the use of **catalysts**.

A catalyst is a substance that increases the rate of a chemical reaction without being used up in the process. A catalyst does take part in the reaction but can be recovered at the end of the reaction unchanged.

Catalysts work by providing a different reaction pathway for the reaction. The reaction rate increases because the new pathway has a lower activation energy than that of the uncatalysed reaction. This can be shown in an energy profile diagram:

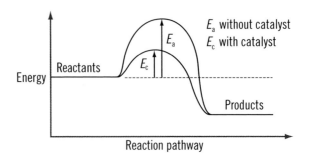

▲ Energy profile diagram with and without catalyst

Because the activation energy of the catalysed reaction is lower, at the same temperature a greater proportion of colliding molecules will achieve the minimum energy needed to react. This can be shown on an energy distribution curve diagram:

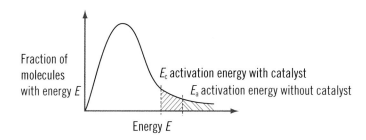

▲ Energy distribution curve with and without catalyst

For a reversible reaction, a catalyst increases the rate of the forward and back reactions by the same amount, therefore it does not affect the position of equilibrium, but the equilibrium is reached more quickly.

There are two classes of catalysts: homogeneous and heterogeneous.

# Homogeneous catalysts

A **homogeneous catalyst** is in the same state (phase) as the reactants. It takes an active part in a reaction rather than being an inactive spectator. Homogeneous catalysis typically involves liquid mixtures or substances in solution.

Examples are:

- Concentrated sulfuric acid in the formation of an ester from a carboxylic acid and an alcohol.

- Aqueous iron(II) ions, $Fe^{2+}(aq)$, in the oxidation of iodide ions, $I^-(aq)$, by peroxodisulfate(VI) ions, $S_2O_8^{2-}(aq)$.

# Heterogeneous catalysts

A **heterogeneous catalyst** is in a different state from the reactants.

Examples are:

- Iron in the Haber process for ammonia production. The reactants, nitrogen and hydrogen, are gases but the catalyst, iron, is a solid.

- Vanadium(V) oxide in the contact process within sulfuric acid manufacture.

- Nickel in the hydrogenation of unsaturated oils in the production of margarine.

Industry relies on catalysts to reduce costs. Since a catalyst speeds up a process by lowering the activation energy of the reaction, less energy is required for the molecules to react, and this saves energy costs. Much of this energy is taken from electricity supplies or by burning fossil fuel so a catalyst also has benefits for the environment. If less fossil fuel is burned, less carbon dioxide will be released during energy production.

Many plastics are formed under high pressures. If a catalyst can be found to enable the reaction to give a good yield at low pressure, the industrial plant will not have to withstand high pressure, and less robust materials can be used in the construction, thus saving money.

**Study point**

A catalyst does not appear as a reactant in the overall equation of a reaction.

**Key terms**

A **homogeneous catalyst** is in the same physical state as the reactants.

A **heterogeneous catalyst** is in a different physical state from the reactants.

**Link**

Hydrogenation of alkenes page 153

**Knowledge check** 6

State what is meant by a catalyst and explain how catalysts work.

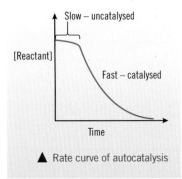
# Enzymes

Enzymes are biological catalysts. They are complex globular proteins which act as homogeneous catalysts in living systems. They usually catalyse specific reactions and work best close to body temperature. Enzyme activity is affected by temperature (it increases until the protein denatures) and pH (different enzymes have differing optimum pH levels). However, enzymes are very effective catalysts, giving a greater increase in reaction rates than inorganic catalysts, e.g. one catalase molecule decomposes 50 000 molecules of hydrogen peroxide per second.

Enzymes are used in a variety of important industrial processes such as food and drink production and the manufacturing of detergents and cleaners. Some examples are:

- Rennin in the dairy industry.
- Yeast and amylase in the brewing industry.
- Lipase and protease in washing powders and detergents.

Some of the benefits are:

- Lower temperatures and pressures can be used, saving energy and costs.
- They operate in mild conditions and do not harm fabrics or food.
- They are biodegradable. Disposing of waste enzymes is no problem.
- They often allow reactions to take place which form pure products, with no side reactions, removing the need for complex separation techniques.

# Studying rates of reaction

To measure the rate of a chemical reaction we need to find a physical or chemical quantity which varies with time. These are some of the main methods (all carried out at constant temperature).

- **Change in gas volume**

  In a reaction in which gas is formed, the volume of the gas can be recorded using a gas syringe at various times:

  e.g. $Mg(s) + 2HCl(aq) \longrightarrow MgCl_2(aq) + H_2(g)$

  or $2H_2O_2(aq) \longrightarrow 2H_2O(l) + O_2(g)$

- **Change in gas pressure**

  Some reactions between gases involve a change in the number of moles of gas. The change in pressure (at constant volume) at various times can be followed using a manometer:

  e.g. $PCl_5(g) \longrightarrow PCl_3(g) + Cl_2(g)$

- **Change in mass**

  If a gas forms in a reaction and is allowed to escape, the change in mass at various times can be followed using weighing scales:

  e.g. $CaCO_3(s) + 2HCl(aq) \longrightarrow CaCl_2(aq) + H_2O(l) + CO_2(g)$

- **Colorimetry**

Some reaction mixtures show a steady change of colour as the reaction proceeds. The concentration of the substance changing colour can be monitored using a colorimeter.

A colorimeter consists of a light source with filters to select a colour of light which is absorbed by the sample. The light passes through the sample and onto a detector or photocell. The photocell, which is connected to a computer, develops an electrical signal proportional to the intensity of the light. The colorimeter is calibrated with solutions of known concentration to establish the relationship between its readings and the concentration of the species being observed.

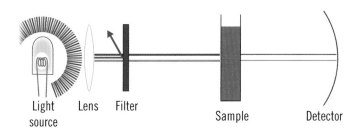

Light    Lens    Filter        Sample       Detector
source

▲ Colorimeter

For example, a colorimeter could be used for the investigation of iodine with propanone $CH_3COCH_3(aq) + I_2(aq) \longrightarrow CH_3COCH_2I(aq) + HI(aq)$ since iodine (brown) is the only coloured species in the reaction.

# Practical activity

A good way of showing how the rate changes during a chemical reaction, as well as illustrating how changing concentration, temperature, particle size or catalysts can affect a chemical reaction, is the **gas collection method**.

For example, reacting magnesium with an acid. A typical method would be:

- Set up the apparatus as shown (ensuring that one reactant is in excess).
- Start the reaction by shaking the magnesium into the acid and start a stopwatch.
- Measure the amount of hydrogen given off at constant intervals (e.g. 30 s).
- Stop the stopwatch when hydrogen is no longer being produced.
- Repeat the experiment with different concentrations of acid / temperature of acid / particle size of magnesium ensuring that all other factors are kept constant.
- Draw a graph of your results.

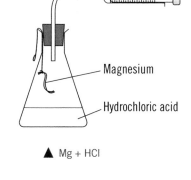

Gas syringe

Magnesium

Hydrochloric acid

▲ Mg + HCl

Analysis of a mixture during the course of a reaction is not always possible or required. To compare rates of reaction under different conditions, a number of experiments may be set up in which initial concentrations of reactants are known and the time taken for each experiment is recorded.

Two examples of this are the 'iodine clock' and precipitation reactions.

**Knowledge check**  **7**

Suggest a method for following the rate of the following reaction:

$2FeCl_3 + 3Na_2CO_3 + 3H_2O$
$\longrightarrow 6NaCl + 2Fe(OH)_3 + 3CO_2$

**≪ Practical check**

Investigation of a rate of reaction by a gas collection method is a **specified practical task.**

Other suitable examples are:

Addition of calcium carbonate to hydrochloric acid. (Sulfuric acid is not suitable.)

Decomposition of hydrogen peroxide. (Different catalysts can be compared.)

Instead of using a gas syringe, the gas can be collected over water using an inverted burette.

**Knowledge check**  **8**

In the reaction between magnesium and hydrochloric acid, why should you not start the reaction by dropping the magnesium into the acid and replacing the bung?

**9** **Knowledge check**

When rates are measured in an iodine clock reaction, state why it is essential that:
(a) All the runs (experiments) are made with the same total volume of liquid.
(b) The temperature does not change during the series of runs.
(c) The peroxide solution is added last.

# Iodine clock reactions

Iodide ions can be oxidised to iodine at a measurable rate. Iodine gives a strongly coloured blue complex with starch solution but if a given amount of thiosulfate ion – with which iodine reacts very rapidly – is added, no blue colour will appear until enough iodine has been formed to react with all the thiosulfate. The time taken for this to occur thus acts as a 'clock' to measure the rate of iodide ions being oxidised.

Oxidising iodide ions by hydrogen peroxide in acid solution is a suitable example of an 'iodine clock' reaction:

$$H_2O_2(aq) + 2H^+(aq) + 2I^-(aq) \xrightarrow{slow} 2H_2O(l) + I_2(aq)$$

$$I_2(aq) + 2S_2O_3^{2-}(aq) \xrightarrow{fast} 2I^-(aq) + S_4O_6^{2-}(aq)$$

By varying the concentrations of the reactants one at a time and measuring the rate, the dependence of rate on concentration for any reactant may be found. A trial run to find what range of concentrations will be suitable should be performed first. The temperature must be kept constant, since rates vary rapidly with changes in temperature.

A typical method for this reaction would be:

- For the trial, add $10.0\,cm^3$ $H_2SO_4$ ($1\,mol\,dm^{-3}$), $10.0\,cm^3$ $Na_2S_2O_3$ ($0.005\,mol\,dm^{-3}$), $15.0\,cm^3$ KI ($0.1\,mol\,dm^{-3}$) and $9.0\,cm^3$ deionised water from burettes into a conical flask.
- Add $1\,cm^3$ of starch solution.
- Measure $5.0\,cm^3$ $H_2O_2$ ($0.1\,mol\,dm^{-3}$) from a burette into a test-tube.
- Rapidly pour the $H_2O_2$ into the flask, simultaneously start a stopwatch and mix thoroughly.
- When the blue colour appears stop the watch.
- Repeat using 5 different concentrations of peroxide, ensuring that the total volume of the mixture is $50\,cm^3$.
- The concentration of peroxide should vary by at least threefold to ensure a good spread of results.

Now rate $\propto \dfrac{1}{time}$ and since total volume is constant in each case, $[H_2O_2] \propto$ volume of peroxide used in each run. Plotting a graph of $\dfrac{1}{time}$ against volume of peroxide will give the relationship between $[H_2O_2]$ and rate.

A similar procedure, but varying the concentration of the potassium iodide, can be used to find the effect of $[I^-]$ on reaction rate. However, for this reaction, the peroxide solution must be added last every time.

# Precipitation reactions

It is not only colour changes in solutions that can be used to follow rates of reaction. Sometimes colourless solutions become more and more cloudy as a precipitate forms and the time taken for a precipitate to form can be used to measure the rate of a reaction.

An example of this is the reaction of thiosulfate ions in acid solution.

$$S_2O_3{}^{2-}(aq) + 2H^+(aq) \longrightarrow S(s) + H_2O(l) + SO_2(g)$$

The reaction is easy to follow since one sulfur atom is formed for each thiosulfate ion reacting and the sulfur makes the reacting solution more cloudy as its concentration increases. By placing the reaction vessel over a black cross (which will disappear from view when a fixed amount of reaction has produced a fixed amount of sulfur) the rates of reaction of solutions of differing concentrations can be compared and the effects of changing concentrations on the rate found.

This is because the time taken for this fixed amount of reaction in all runs is inversely proportional to the rate of reaction, i.e. if the reaction is fast the cross will disappear quickly and vice versa.

A trial run to find what range of concentrations will be suitable should be performed first. The temperature must be kept constant, since rates vary rapidly with changes in temperature.

A typical method for this reaction would be:

- For the trial add $6.0\,cm^3$ $Na_2S_2O_3$ $(1\,mol\,dm^{-3})$ and $4.0\,cm^3$ water into a conical flask.
- Add $10.0\,cm^3$ $HNO_3$ $(0.1\,mol\,dm^{-3})$ from a burette into a test tube.
- Rapidly pour the acid into the flask, simultaneously start a stopwatch and mix thoroughly.
- Place the flask over a black cross.
- Stop the watch the moment the black cross can no longer be seen.
- Repeat the procedure with three different concentrations of $Na_2S_2O_3$ keeping the volume of acid at $10.0\,cm^3$ and the total volume of the mixture at $20.0\,cm^3$. (Ensure that there is a good spread of concentration.)
- Repeat the procedure with three different concentrations of $HNO_3$ keeping the volume of thiosulfate at $6.0\,cm^3$ and the total volume of the mixture at $20.0\,cm^3$. (Ensure that there is a good spread of concentration.)

Since rate $\propto \dfrac{1}{time}$ plotting a graph of $\dfrac{1}{time}$ against concentration of thiosulfate at constant acid concentration will give the relationship between $[S_2O_3{}^{2-}]$ and rate.

Plotting a graph of $\dfrac{1}{time}$ against concentration of acid at constant thiosulfate concentration will give the relationship between $[HNO_3]$ and rate.

**≪ Practical check**

You should ensure that you obtain sensible reaction times. Do not use concentrations that give a time of under 20 s because the error in measuring the time will be large.

# Test yourself

1. When marble chips react with hydrochloric acid the following reaction occurs

$$CaCO_3(s) + 2HCl(aq) \longrightarrow CaCl_2(aq) + H_2O(l) + CO_2(g)$$

   (a) Suggest two methods for measuring the rate of this reaction. [2]

   (b) State what could be done to the marble chips to increase the rate of this reaction. [1]

2. (a) The diagram below shows the distribution of molecular energies for a sample of gas.

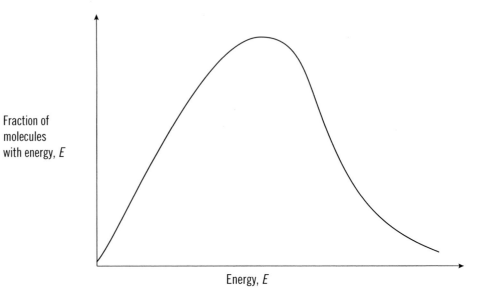

Fraction of
molecules
with energy, $E$

Energy, $E$

     (i) Use this diagram to explain how a catalyst increases the rate of reaction. [2]

     (ii) On the diagram draw the distribution curve of molecular energies for the same sample of gas at a lower temperature. [1]

   (b) Magnesium reacts with sulfuric acid to give magnesium sulfate and hydrogen.

$$Mg + H_2SO_4 \longrightarrow MgSO_4 + H_2$$

The table below shows the initial concentration of sulfuric acid and its concentration after 60 seconds.

| Concentration of sulfuric acid / mol dm$^{-3}$ | Time / s |
|---|---|
| 0.500 | 0 |
| 0.242 | 60 |

     (i) Use the values to calculate the average rate of reaction over this time interval. State the unit. [2]

     (ii) State whether the initial rate of reaction would be the same as the average rate. Give a reason for your answer. [2]

     (iii) The reaction between magnesium and sulfuric acid is an exothermic process.

Sketch an energy profile for this reaction, clearly showing the activation energy, $E_a$. [2]

3. One of the stages during the Contact process is the conversion of sulfur dioxide to sulfur trioxide:

$$2SO_2(g) + O_2(g) \rightleftharpoons 2SO_3(g) \qquad \Delta H = -197 \text{ kJmol}^{-1}$$

State, giving reasons, what happens to the rate of the reaction as

(a) the temperature is increased, [3]

(b) the total pressure is increased. [3]

4. Explain why so many catalysts are used in chemical industrial processes. [3]

5. The following results were obtained in an experiment to measure the rate of oxidation of iodide ions by hydrogen peroxide in acid solution as shown in the equation. The reaction was carried out at a temperature of 20 °C.

$$H_2O_2 + 2H^+ + 2I^- \longrightarrow I_2(\text{brown}) + 2H_2O$$

| Time/ s | 0 | 100 | 200 | 300 | 400 | 500 |
|---|---|---|---|---|---|---|
| $I_2$ concentration/ mol dm$^{-3}$ | 0 | 0.0115 | 0.0228 | 0.0347 | 0.0420 | 0.0509 |

(a) Plot these results on graph paper, labelling the axes and selecting a suitable scale. Draw the line of best fit. [3]

(b) Use the graph to calculate the initial rate of reaction and give the units. [3]

(c) Describe briefly the key features of the method that would have been used to obtain these results. [3]

(d) A similar experiment was carried out using hydrogen peroxide and iodide solutions of different concentrations. The initial rates calculated for each reaction are shown in the table.

| Concentration of $H_2O_2$ / relative units | Concentration of $I^-$ / relative units | Initial rate / relative units |
|---|---|---|
| 0.60 | 0.050 | $4.1 \times 10^{-4}$ |
| 1.2 | 0.050 | $8.0 \times 10^{-4}$ |
| 1.2 | 0.10 | $1.6 \times 10^{-3}$ |

Analyse the data and state the relationship between the concentrations of hydrogen peroxide and iodide ions and the initial rate of reaction. [2]

# The wider impact of chemistry

The large-scale production of chemicals and combustion of fossil fuels is causing serious problems, especially in the rise of $CO_2$ levels and its effect on the climate. Efforts are being made to use fewer fossil fuels or to use alternative fuels and to use fuels more efficiently.

Green chemistry addresses these problems by attempting to prevent waste, use safer and more efficient methods, increase atom economy and reduce pollution.

## Topic contents

You should be able to demonstrate and apply knowledge and understanding of the following:

- Why non-renewable fossil fuels are being replaced.
- Alternatives to non-renewable fossil fuels.
- Carbon neutrality and how it may be achieved.
- The role of green chemistry in improving sustainability in all aspects of chemical synthesis and energy production.

# Social, economic and environmental impact

Both the chemical industry itself, and obtaining the materials for chemical processes, are major sources of employment and many towns have grown up around such centres. Industries have been sited near sources of raw materials (such as iron ore, coal, water) and good transport. There is now a greater realisation of the importance of providing a healthy and safe environment for workers and their families living near the factories.

Economically, the requirements today are to produce good quality products efficiently and safely with a well-paid and fulfilled workforce.

▲ Environmentally-friendly chemical works in Japan

## The energy problem

Energy production is a major issue affecting not only the chemical industry but all aspects of modern life. Huge and increasing amounts of energy are needed by the world's expanding and developing population for industry, transport, electricity generation and domestic heating. Serious problems of supply lie ahead in that the finite and non-renewable resources of fossil fuels (coal, oil and gas) are being rapidly depleted.

Non-renewable fossil fuels are being replaced by nuclear energy and renewable sources such as wind and water power in its various forms, combustion of biomass, geothermal energy and solar energy.

Hydrogen is spoken of as a future fuel but does not exist naturally on Earth and must be prepared at a cost of energy.

▲ Safer working conditions for employees

### Carbon neutrality

Underlying the energy problem is the knowledge that a huge amount of carbon dioxide is being passed into the atmosphere through fuel combustion and other industrial processes, such as making concrete. The carbon dioxide concentration has risen from around 300 to 400ppm (parts per million) in the last 100 years, and there seems little doubt that this has led to global warming through an increase in the greenhouse effect. The consequences of this warming may be extremely serious.

**Carbon neutrality** is achieved with combustible fuel sources by replanting trees and sugar cane, etc., to match those being burnt, so that the $CO_2$ consumed in their photosynthesis equals or exceeds that generated by their combustion.

▲ Wind power

### Biomass energy

A renewable source that has carbon neutrality is the combustion of biomass.

Biomass is plant or animal material used for energy production. Burning plant-derived biomass releases $CO_2$, but photosynthesis cycles the $CO_2$ back into new crops. In some cases, this recycling of $CO_2$ from plants to atmosphere and back into plants can even be $CO_2$ negative, as a relatively large portion of the $CO_2$ is moved to the soil during each cycle.

The essential processes are shown in the simple equation:

$$CH_2O + O_2 \underset{\text{photosynthesis}}{\overset{\text{combustion}}{\rightleftharpoons}} CO_2 + H_2O$$

sugar

>> **Key term**

**Carbon neutrality** means that a chemical process such as fuel combustion does not lead to an overall increase in $CO_2$ levels. Although the combustion does generate $CO_2$ this is offset by the fact that the fuel has absorbed the same amount of $CO_2$ in being made by photosynthesis.

**Knowledge check 1**

Which of the following combustion fuels is carbon neutral?

Coal – Oil – Methane – Hydrogen – Liquefied petroleum gas (LPG)

▲ Solar power

▲ Nuclear power

**‹ Link ›**

Fuels page 143

**‹ Link ›**

Atom economy page 44

**‹ Link ›**

Catalysts page 116

▲ Jet engine

## Solar power

A renewable source that generates electricity directly is solar power.

Solar power refers to capturing the energy from the Sun and subsequently converting it into electricity. Sunlight may be converted directly into electricity by solar or photovoltaic cells. The energy efficiency of these cells has improved over time and in 2019 the average efficiency was 18%.

The total amount of solar energy available on Earth is vastly in excess of the world's current and anticipated energy requirements. Unfortunately, though solar energy itself is free, the high cost of its collection, conversion, and storage still limits its exploitation in many places.

## Nuclear power

A non-renewable source that generates electricity directly is nuclear power.

Around 20% of the electricity needed may be produced in this way by the neutron-aided fission of uranium in which less than 1% of mass is converted to energy through the equation $E = mc^2$. The method has been in use since the 1970s but has drawbacks with regard to radioactive emissions and the safe disposal of radioactive wastes.

A better method on paper is in the fusion of hydrogen to form helium, as in the Sun, which converts a higher percentage of mass into energy with fewer problems of radioactive products. Unfortunately, despite several decades of research, it has not yet been possible to solve the associated technical problems.

# The role of green chemistry

The aim of green chemistry is to make the chemicals and products that we need with as little impact on the environment as possible. This means:

### Renewable raw materials
Using renewable raw materials such as plant-based compounds whenever possible.

### Saving energy
Using as little energy as possible and getting this from renewable sources such as biomass, solar, wind and water rather than from finite fossil fuels such as oil, gas and coal.

### High atom economy
Using methods having high atom economy so that a high percentage of the mass of reactants ends up in the product, giving little waste.

### Catalysts
Developing better catalysts and biocatalysts, such as enzymes, to carry out reactions at lower temperatures and pressures to save energy and avoid high-pressure plants.

### Toxic materials and solvents
Avoiding the use of toxic materials if possible and ensuring that no undesirable co-products or by-products are released into the environment.

Avoiding the use of solvents, especially volatile organic compounds (VOCs) that are bad for the environment. Diesel engines liberate toxic nitrogen oxides ($NO_x$) into the atmosphere. Addition of ammonia converts these into harmless nitrogen. (See final bullet point on the next page.)

### Biodegradable products
Making products that are biodegradable at the end of their useful lives, where possible.

## Summary of green principles

1 Prevent waste
2 Increase atom economy
3 Use safer methods, chemicals and solvents
4 Increase energy efficiency
5 Use renewable raw materials (feedstocks)
6 Use catalysts
7 Prevent pollution and accidents
8 Design for biodegradation.

▲ Biomethane

Some examples involving the principles above:

- Anaerobic digestion – methanogenic bacteria act on manures and vegetable wastes in an oxygen-free atmosphere to produce biogas containing 70% methane for use in electricity generation and leave a fertiliser as a residue.

- Feedstocks use renewable raw materials such as palm oil for making biodiesel.

- Using liquid waste carbon dioxide under pressure (supercritical) as a residue-free solvent, e.g. in decaffeinating coffee beans, dry cleaning and blowing foam plastics.

- Simvastatin (a cholesterol-reducing drug) is made from cheap raw materials using an engineered enzyme made from *E. coli*. Avoids multi-step synthesis using toxic reagents and gives higher yield.

- 1,3-propanediol made from corn syrup using genetically modified *E. coli* uses 40% less energy and emits 20% less greenhouse gas than earlier processes. 1,3-propanediol can also be made from glycerol waste in biodiesel production using bacteria. It is used in making polyesters.

- Prevent ing pollution by toxic nitrogen oxides ($NO_x$) in car/diesel exhausts by adding ammonia to give the redox reaction:

▲ African oil palms

⟨ **Link** ⟩

Oxidation numbers pages 13, 67

$$4NH_3 + 4NO + O_2 \rightleftharpoons 4N_2 + 6H_2O$$

The use of palm oil is an example of conflicting interest in green chemistry.

Palm oil is made from the fruits of trees called African oil palms. It is an extremely versatile oil that has many different properties and applications, which makes it widely used. In 2018, one-half of Europe's palm oil imports were used for biodiesel. However, palm oil is also in around half of the packaged products we find in supermarkets, everything from pizza, doughnuts and chocolate, to deodorant, shampoo, toothpaste and lipstick.

Palm oil is an incredibly efficient crop, producing more oil per land area than any other equivalent vegetable oil crop. Globally, palm oil supplies 35% of the world's vegetable oil demand on just 10% of the land. To get the same amount of alternative oils such as soybean or coconut oil you would need anything between 4 and 10 times more land.

Unfortunately, palm oil has been and continues to be a major driver of deforestation of some of the world's most biodiverse forests, destroying the habitat of already endangered species. Huge tracts of rainforest are being bulldozed or torched, sometimes illegally, to make room for more plantations. This forest loss, coupled with conversion of carbon-rich peat soils, results in millions of tonnes of greenhouse gases being released into the atmosphere and contributing to climate change.

Palm oil can be produced more sustainably and things can change. The Roundtable on Sustainable Palm Oil was formed in 2004 in response to increasing concerns about the impacts palm oil was having on the environment and on society.

In 2012, the UK government recognised that the UK was part of the palm oil problem and could also be part of the solution. They set a commitment for 100% of the palm oil used in the UK to be from sustainable sources that don't harm nature or people. In 2016, 75% of the total palm oil imports to the UK were sustainable. In 2018 the EU stated that it plans to phase out palm oil from transport fuel.

# Test yourself

1. Liquefied petroleum gas or LPG is a mixture of hydrocarbon gases used as fuel in heating appliances and cooking equipment. The main components are propane and butane.

   (a) The equation for the complete combustion of propane is

   $$C_3H_8 + 5O_2 \longrightarrow 3CO_2 + 4H_2O$$

   Write the equation for the complete combustion of butane. [1]

   (b) Calculate the energies produced **per gram** when propane and butane are combusted.

   $\Delta_cH$ propane = −2200 kJ mol$^{-1}$        $\Delta_cH$ butane = −2811 kJ mol$^{-1}$   [2]

   (c) Calculate the mass of carbon dioxide formed when
   (i)  one gram of propane
   (ii) one gram of butane
   is combusted. [2]

   (d) Use your answers to (b) and (c) to state whether propane or butane is the more environmentally friendly fuel. Give a reason for your answer. [2]

2. Motor vehicle environmental effects are judged on the grams of carbon dioxide emitted per kilometre of travel.

   The combustion equations for bioethanol and ethane and the energy liberated are:

   $$C_2H_5OH + 3O_2 \longrightarrow 2CO_2 + 3H_2O \qquad \Delta H = -1371 \text{ kJ mol}^{-1}$$
   $$C_2H_6 + 3\tfrac{1}{2}O_2 \longrightarrow 2CO_2 + 3H_2O \qquad \Delta H = -1560 \text{ kJ mol}^{-1}$$

   Bearing in mind that the car is driven by the energy liberated, state which fuel would be the more environmentally friendly per mol on the basis of these results. Give a reason for your answer.

   There is, however, another factor to be considered. Discuss this and state whether or not your initial conclusion should be altered. [3]

3. The vast majority of motor vehicles worldwide are powered by petrol or diesel. Many vehicle manufacturers around the world have made the development of alternative fuels a priority.

   Give **one** advantage and **one** disadvantage of using
   (a) hydrogen [2]
   (b) bioethanol [2]
   Two fuels under consideration are hydrogen and bioethanol as vehicle fuel.

4. Many people are very concerned about the current levels of carbon dioxide in the atmosphere.

   The table below shows the carbon dioxide levels in the atmosphere in certain years.

   | | Year | | | |
   |---|---|---|---|---|
   | | 1950 | 1980 | 2000 | 2018 |
   | Carbon dioxide levels in the atmosphere / ppm | 310 | 339 | 368 | 407 |

   (a) Give **two** reasons for the increase in carbon dioxide levels between 1950 and 1980. [2]

   (b) Calculate the percentage increase in carbon dioxide levels between 2000 and 2018. [1]

   (c) The increase in carbon dioxide levels from 1950 to 1980 and 1980 to 2000 is 29 ppm.

   A student said that since the increase was the same in both periods, this shows that there was a steady increase in carbon dioxide levels between 1950 and 2000.

   Give **two** reasons why the student is incorrect. [2]

# Organic compounds

Organic compounds are hugely important in everyday life, in understanding the reactions within living species and in industrial processes. To allow new compounds to be synthesised, and their properties to be linked to potential uses, it is vital that the patterns of behaviour, linked to the structure of the molecules, are recognised. These patterns include the nature of the reactions that the compounds undergo.

International agreement on how compounds are named ensures that scientists throughout the world recognise the compound being described.

You should be able to demonstrate and apply your knowledge and understanding of:

- How to represent simple organic compounds using shortened, displayed and skeletal formulae.
- Nomenclature rules relating to alkanes, alkenes, halogenoalkanes, alcohols and carboxylic acids.
- The effect of increasing chain length and the presence of functional groups on melting/boiling temperature and solubility.
- The concept of structural isomerism.

## Maths skills »

- Use ratios to write formulae and construct and balance equations.
- Use ratios, fractions and percentages to determine empirical and molecular formulae.
- Visualise and represent 2D and 3D forms including two-dimensional representations of 3D objects.
- Use standard and ordinary form, decimal places and significant figures (this point is covered in chapter 2.5 – see page 142).

## Topic contents

# Organic compounds

The term organic is used for compounds of carbon in general. There are many thousands of such compounds and whole textbooks have been written about them.

All organic compounds contain carbon and hydrogen and some also contain oxygen, nitrogen, sulfur, phosphorus and the halogens. The elements and their arrangement in a compound define the **functional group** and the functional group defines to which **homologous series** the compound belongs.

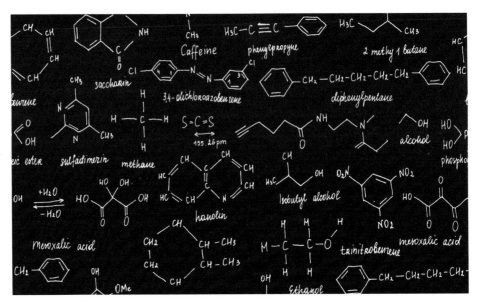

▲ Organic compounds

# Naming organic compounds

Since there are so many organic compounds it is essential that a system exists that gives each one a name on which all chemists agree. The international system used is the IUPAC system. To name a compound you have to know the homologous series to which it belongs. The homologous series in this section are:

Alkanes – **saturated hydrocarbons** with no C to C multiple bonds

Alkenes – **unsaturated** hydrocarbons with a C to C double bond

Halogenoalkanes – compounds in which one or more hydrogens in an alkane have been replaced by a halogen

Alcohols – compounds containing –OH as the functional group

Carboxylic acids – compounds containing –COOH as the functional group

You also need to know the stem that applies to the number of carbon atoms.

| Number of carbons | Stem | | Number of carbons | Stem |
|---|---|---|---|---|
| 1 | meth | | 6 | hex |
| 2 | eth | | 7 | hept |
| 3 | prop | | 8 | oct |
| 4 | but | | 9 | non |
| 5 | pent | | 10 | dec |

# Rules for naming organic compounds

Compound names have the format of prefix-stem-suffix. For example, 2-methylbutane has the prefix '2-methyl', the stem 'but' and the suffix 'ane'. The following rules will help you in naming compounds correctly.

1  Find the longest carbon chain. Using the table on the previous page, this forms the stem of the name.

2  Number the C atoms in the chain, starting from the end that gives any side chains or functional groups the smallest number possible. Some functional groups, e.g. acids, always start as carbon number 1.

3  Identify all the side chains and substituted functional groups using the appropriate prefixes, e.g. methyl or chloro. If there is more than one side chain or substituted group the same, use di for 2, tri for 3 and tetra for 4.

4  When writing the full name, list prefixes in alphabetical order, e.g. ethyl comes before methyl.

5  Common prefixes are methyl for a $-CH_3$ group, ethyl for a $-CH_2CH_3$ group, chloro- for an attached chlorine, bromo- for an attached bromine, hydroxy- for an attached OH group (used when another functional group is present that has to be at the end of the name).

6  A saturated hydrocarbon is shown by -ane, a C=C is shown by -ene, an $-OH$ is shown by $-ol$, a halogen is shown by halogeno and a $-COOH$ is shown by -oic acid.

There is no need to be able to state these rules but you do need to be able to apply them to naming particular compounds.

## Examples of how to name compounds

### Example 1

```
    H   H   H   H
    |   |   |   |
H — C — C — C — C — H
    |   |   |   |
    H   |   H   H
    H — C — H
        |
        H
```

The longest carbon chain has 4 carbons and so the name is based on but.

It is saturated so the name ends in -ane.

Numbering from end to give lowest number for side chain gives $-CH_3$ on carbon number 2.

Name is **2-methylbutane**.

### Example 2

```
        H
        |
    H — C — H
    H   |   H
    |   |   |
H — C — C — C — Cl
    |   |   |
    H   |   H
    H — C — H
        |
        H
```

It does not matter which way you number, there are still 3 carbons in the longest chain so the name is based on prop.

There are 2 $-CH_3$ groups on number 2 carbon and a chlorine on number 1 carbon.

Name is **1-chloro-2,2-dimethylpropane**.

**Exam tip**

Systematic names for compounds include commas and hyphens. You must include enough of these to make your name unambiguous. For example, 22-dimethyl propane is not the same as 2,2-dimethylpropane.

**Knowledge check** 1

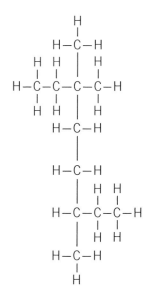

Find and name the longest carbon chain in the compound above.

**2 Knowledge check**

Draw the formulae of:

(a) 4-chlorobut-1-ene

(b) 2,2-dimethylpentan-3-ol

**3 Knowledge check**

What are the names of the following?

(a)

(b)

## Example 3

There are 5 carbons in the longest carbon chain and numbering from the left-hand end gives side chain and functional group the lowest numbers. Name based on pent.

Presence of C=C (alk**ene**) shown by ene (on C 1) and OH (alcoh**ol**) shown by ol (on C 3).

Name is **pent-1-en-3-ol**

## Example 4

There are five carbons in the longest chain – whether you count left to right or along the bent chain. Name based on pent.

Presence of –COOH shown by oic acid (this C is C 1) and –$CH_2CH_3$ on C 3.

Name is **3-ethylpentanoic acid**.

**Key terms**

**Molecular formula** shows the atoms, and how many of each type there are, in a molecule of a compound.

**Displayed formula** shows **all** the bonds and atoms in the molecule.

**Shortened formula** shows the groups in sufficient detail that the structure is unambiguous.

**Skeletal formula** shows the carbon/ hydrogen backbone of the molecule as a series of bonds with any functional groups attached.

**‹Link›**

Empirical formulae are introduced in Section 1.3 and discussed further on page 134

# Types of formulae

The formula of a particular compound can be shown in several ways. The way chosen usually depends on the use being made of the formula.

The **molecular formula** shows the atoms, and how many of each type of atom there are, in a molecule of a compound. All the atoms of the same element are drawn together and therefore the functional group is not always obviously recognisable.

A **displayed formula** shows all the atoms, and the bonds linking them, in a compound. This clearly shows the functional group present and would be used when considering the mechanisms of organic reactions.

The **shortened formula** shows the functional group and structure in sufficient detail so that the compound is unambiguous. It cannot be used if details of the bonds are needed.

The **skeletal formula** shows the functional groups without the distraction of unreactive chains. It can reduce confusion when complex molecules are being considered and is widely used in research work.

## Examples

**1** Using 1–chloro–2,2–dimethylpropane

Molecular formula: $C_5H_{11}Cl$

Displayed formula:

$$
\begin{array}{c}
\text{H} \\
| \\
\text{H}-\text{C}-\text{H} \\
\text{H} \quad | \quad \text{H} \\
| \quad | \quad | \\
\text{H}-\text{C}-\text{C}-\text{C}-\text{Cl} \\
| \quad | \quad | \\
\text{H} \quad | \quad \text{H} \\
\text{H}-\text{C}-\text{H} \\
| \\
\text{H}
\end{array}
$$

Shortened formula: $CH_3C(CH_3)_2CH_2Cl$

Skeletal formula:

**2** Using pent-1-en-3-ol

Molecular formula: $C_5H_{10}O$

Displayed formula:

$$
\begin{array}{c}
\text{O-H H} \quad \text{H} \\
| \quad | \quad | \\
\overset{\text{H}}{\underset{\text{H}}{>}}\text{C}=\text{C}-\text{C}-\text{C}-\text{C}-\text{H} \\
| \quad | \quad | \quad | \\
\text{H} \quad \text{H} \quad \text{H} \quad \text{H}
\end{array}
$$

Shortened formula: $CH_2CHCH(OH)CH_2CH_3$

Skeletal formula:

**≪ Exam tip**

In an exam, make sure that you use the type of formula that the question asks for – molecular, displayed, shortened or skeletal. There could also be empirical formulae – these are covered later!

**Knowledge check  4**

The shortened formula of a compound is $CH_3CCl(OH)CHCH_2$.

(a) What is its molecular formula?

(b) Draw its displayed formula.

(c) Draw its skeletal formula.

**≫ Study point**

In a skeletal formula, carbon atoms are never shown and only hydrogen atoms within the functional group are shown.

# Homologous series

As described earlier, each compound belongs to a particular homologous series. An homologous series is a set of compounds that:

**1** Can be represented by a general formula.

**2** Differ from their neighbour in the series by $CH_2$.

**3** Have the same functional groups and so very similar chemical properties.

**4** Have physical properties that vary as the $M_r$ of the compound varies.

The homologous series in this section are alkanes (saturated hydrocarbons), alkenes (unsaturated; functional group C=C), alcohols (functional group –OH), halogenoalkanes (functional group –F, –Cl, –Br, –I) and carboxylic acids (functional group –COOH).

**5 Knowledge check**

Complete the following sentence.

Alkenes have similar ............. properties but show a trend in ............ properties because they all belong to the same ............. .

**6 Knowledge check**

What is the molecular formula of the alkane with 100 carbon atoms in each molecule?

**Key term**

**The empirical formula** is the formula of a compound with the atoms of the elements in their simplest integer ratio.

---

An example of a general formula is shown using the alkanes. Using displayed formulae the series is:

H
|
H—C—H
|
H

H  H
|  |
H—C—C—H
|  |
H  H

H  H  H
|  |  |
H—C—C—C—H
|  |  |
H  H  H

This gives molecular formulae of $CH_4$, $C_2H_6$, $C_3H_8$.

This means that the general formula is $C_nH_{2n+2}$ where $n$ is an integer. For example, if $n = 3$ that means that $2n + 2 = 8$ so formula is $C_3H_8$.

The effects of the functional group and the way in which physical properties, particularly meting temperature, boiling temperature and solubility in water, vary within an homologous series are considered later.

# Empirical formulae

The **empirical formula** shows the formula of a compound with the atoms of the elements in their simplest ratio. This may be the molecular formula but the molecular formula could also be any multiple of the empirical formula.

**Example**

Molecular formula of ethane $= C_2H_6$

Ratio C:H in molecular formula $= 2:6$

Simplest ratio C:H $= 1:3$

Empirical formula of ethane $= CH_3$

### Determining an empirical formula

For organic substances, most experimental methods are based on burning a known mass of the compound in excess oxygen and measuring the masses of water and carbon dioxide produced.

**Example**

We can represent the combustion of a hydrocarbon in excess oxygen by the following equation:

$$C_xH_y + \text{excess } O_2 \longrightarrow xCO_2 + \frac{y}{2}H_2O$$

This means if we know the masses of $CO_2$ and $H_2O$ produced, we can calculate the number of moles, $x$ and $y$, and these values can be used to help deduce the empirical formula of the hydrocarbon.

$$x = \frac{\text{mass } CO_2}{44} \qquad y = \left(\frac{\text{mass } H_2O}{18.02}\right) \times 2$$

Divide through by the smallest number to convert $x$ and $y$ into integer (whole number) values.

This allows the mass of carbon and hydrogen to be calculated but not other elements present – this is considered in the calculation below. You do not need to know how the halogen content of a compound is determined.

Sometimes questions quote the percentages of elements present. These percentages can be used as equivalent to mass since they would be the mass of the element in 100g of compound. The ratio of the atoms of each element is then found using the formula:

$$\text{number of moles} = \frac{\text{mass}}{A_r}.$$

## Worked example

A compound has the following percentage composition: C = 12.78%; H = 2.15%; Br = 85.07%. Find its empirical formula.

### Answer

Ratio C:H:Br $= \dfrac{12.78}{12} : \dfrac{2.15}{1.01} : \dfrac{85.07}{79.9} = 1.07 : 2.13 : 1.06$

To find the ratio as integers, divide by the smallest number – in this case 1.06.

Ratio C:H:Br $= 1:2:1$.

Empirical formula is $CH_2Br$.

This calculation **could** give the molecular formula, although not in this case since no such compound is possible. The molecular formula must therefore be a multiple of this empirical formula and you need to find which multiple. For this, the $M_r$ of the compound is needed.

In this case, assume the $M_r$ is 188.

$M_r$ of empirical formula $= 94$ so the molecular formula must be twice the empirical formula.

Molecular formula $= C_2H_4Br_2$

Sometimes, however, a calculation quotes the experimental data and in this case you need to calculate the percentages of the elements from these.

## Worked example

Compound **A** contains carbon, hydrogen and oxygen. 1.55 g of compound A was burned in excess oxygen and 1.86 g of water and 3.41 g of carbon dioxide were formed.

Calculate the empirical formula of **A**.

*There are different ways of answering this question – two different worked answers are shown below.*

### Answer 1

Compound $A = C_xH_yO_z$

The combustion of A in excess oxygen can be represented by:

$C_xH_yO_z + \text{excess } O_2 \longrightarrow xCO_2 + \dfrac{y}{2}H_2O$

1.55 g $\qquad\qquad$ 3.41 g $\quad$ 1.86 g

$x = \dfrac{3.41}{44} = 0.0775 \text{ mol}; \quad y = \left(\dfrac{1.86}{18.02}\right) \times 2 = 0.206 \text{ mol}$

To find z we first find the mass of oxygen in the original sample by subtracting the mass of carbon and the mass of hydrogen:

Mass of oxygen in A $= 1.55 - (0.0775 \times 12) - (0.206 \times 1.01) = 0.412$ g

Therefore $\quad z = \dfrac{0.412}{16} = 0.0257$ mol

Empirical formula $= C_{0.0775}H_{0.206}O_{0.0257}$

Divide these numbers by 0.0257 to get the answer 3 : 8 : 1

So the empirical formula is $C_3H_8O$.

*Continued on next page* ▶

A compound has a $M_r$ of approximately 55. Its percentage composition is C = 40.00%, H = 6.67% and O = 53.33%. Find its empirical formula and hence its molecular formula.

**◀◀ Exam tip**

Always show clearly that you have calculated the $M_r$ for your empirical formula. You do not need to know the exact $M_r$ of the compound as it has to be a multiple of the empirical $M_r$.

**◀◀ Maths tip**

When you divide by the smallest number to give a whole number ratio, do not cheat! If the ratio is, for example, 1 : 1.5 this is 2 : 3. If the ratio is 1 : 1.67 this is 3 : 5. If your ratio does not simplify to give whole numbers, go back and check – you have made a mistake! One common mistake is to round numbers too soon. Don't do this until the end.

## 8 Knowledge check

A sample of a hydrocarbon was burned completely in oxygen. 0.660 g of carbon dioxide and 0.225 g of water were formed. The $M_r$ of the compound was approximately 80. Find its empirical formula and hence its molecular formula.

### Link

The $M_r$ of a compound can be found using mass spectrometry. See page 177

*Answer 2*

*An alternative approach to this calculation is to find the percentage of each element.*

To find the % of carbon:

In $CO_2$ $\dfrac{12}{44}$ of the mass is the mass of carbon

3.41 g of $CO_2$ contains $3.41 \times \dfrac{12}{44}$ g of C $= 0.93$ g

% C $= \dfrac{0.93}{1.55} \times 100 = 60.0\%$

To find the percentage of hydrogen:

In $H_2O$ $\dfrac{2}{18.02}$ of the mass is the mass of hydrogen

1.86 g of $H_2O$ contains $1.86 \times \dfrac{2}{18.02} = 0.206$ g

% H $= \dfrac{0.206}{1.55} \times 100 = 13.3\%$

To find the percentage of oxygen:

Total % C and H $= 73.3\%$. Rest is oxygen so % O $= 26.7\%$

Ratio C : H : O $= \dfrac{60.0}{12} : \dfrac{13.3}{1} : \dfrac{26.7}{16} = 5.0 : 13.3 : 1.67$

Divide by smallest C : H : O $= 3 : 8 : 1$

Empirical formula $= C_3H_8O$

## Worked example

The $M_r$ of **A** is approximately 65. What is the molecular formula of **A**?

*Answer*

$M_r$ of empirical formula $= 36 + 8 + 16 = 60$.

Molecular formula of **A** $= C_3H_8O$

# Isomerism

There are two types of isomerism – structural and *E-Z*. *E-Z* is covered in the next chapter.

**Structural isomers** are compounds with the same molecular formula but different structural formulae, i.e. arrangement of the atoms. Structural isomerism can arise in several ways.

## 1 Chain isomerism

This occurs when the carbon chain of the molecule is arranged differently. Usually one isomer has a straight chain and others have branched chains.

$CH_3CH_2CH_2CH_3$ butane    and

$$CH_3{-}\overset{\displaystyle CH_3}{\underset{\displaystyle H}{\overset{|}{\underset{|}{C}}}}{-}CH_3$$    2-methylpropane

## 2 Position isomerism

This occurs when the functional group is in a different position in the molecule.

$CH_2ClCH_2CH_3$ 1-chloropropane and $CH_3CHClCH_3$ 2-chloropropane

## 3 Functional group isomerism

This occurs when the functional group in the compounds is different, e.g. $C_3H_8O$.

$CH_3CH_2OCH_3$ methoxyethane (an ether)

and

$CH_3CH_2CH_2OH$ propan-2-ol (an alcohol)

You need to be able to recognise structural isomers but do not need to be able to classify them into chain, position or functional group.

As the molecular formula increases in size, many isomers are possible. $C_5H_{12}O$ has many isomers some of which are shown below. You should be able to draw these isomers but you are not expected to state which type of structural isomerism each one shows.

> ## Key term
>
> **Structural isomers** are compounds with the same molecular formula but with different structural formulae.

> ## Study point
>
> You will be expected to recognise, name and draw isomers of all the homologous series in this section.

> ## Knowledge check 9
>
> (a) Draw displayed formulae for structural isomers of $C_5H_{12}$. Name the isomers.
>
> (b) Draw skeletal formulae for two straight chain structural isomers of $C_5H_{10}$ that involve different positions of the double bond. Name the isomers.

> ## Knowledge check 10
>
> Which of the compounds A, B and C are isomers of each other?
>
>

> ## Link
>
> For *E–Z* isomerism, see p149

# Melting and boiling temperatures

## The effect of chain length on melting and boiling temperatures

In a solid, the particles are held together rigidly and they can only vibrate about a fixed position. When the solid melts, the forces that hold it rigidly have to be overcome. Although there is less order in a liquid and the particles are further apart, significant attractive forces are still present and these have to be overcome when the liquid changes into a gas. This means that energy is needed to overcome forces whenever a substance melts or boils. This energy is generally in the form of heat. When comparing substances, the one that has a higher melting or boiling temperature can be predicted by looking at the strength of these forces.

Hydrocarbons are simple covalent molecules consisting of only carbon and hydrogen. Since the electronegativities of carbon and hydrogen are similar, hydrocarbons are non-polar. This means that only temporary dipole–temporary dipole **Van der Waals forces** are present between the molecules. These are weak intermolecular forces.

Intermolecular temporary dipole–temporary dipole forces act between the surfaces of the molecules. The more surface there is in contact, the stronger the forces. This means that more energy is needed to overcome the forces and the melting and boiling temperatures are higher.

The graph below shows how the boiling temperature of alkanes increases as the chain length increases.

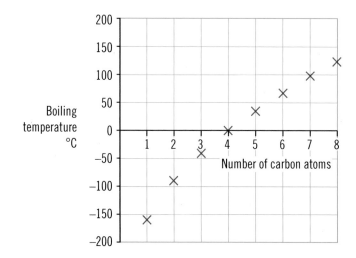

Small hydrocarbons are gases at room temperature, larger ones are liquids and even larger ones are solids.

### Key term

**Van der Waals forces** are dipole–dipole or temporary dipole–temporary dipole interactions between atoms and molecules.

### Link

Forces between molecules page 53

### Study point

Make sure you understand the difference between **inter**molecular and **intra**molecular forces. Intermolecular are between molecules and intramolecular forces are within the same molecule.

### 11 Knowledge check

Draw **two** molecules of water showing all the bonds. Mark on this diagram an intermolecular force and an intramolecular force.

# The effect of branching on boiling temperatures

Boiling temperatures of hydrocarbons are not only affected by chain length, they are also affected by the shape of the molecule. Different structural isomers will have different boiling temperatures because they have different surface areas that can be in contact. Consider pentane and its isomer 2,2-dimethylpropane.

Pentane
Boiling temperature 36 °C

2,2-Dimethylpropane
Boiling temperature 10 °C

Straight chains – more
surface area contact

Many branches – little
surface contact

Looking at the shapes of these molecules, the more branches that are present the more spherical the molecule. When spheres are packed together the surface area available for contact is very small, whereas 'sausage-shaped' molecules have much more surface area available. This means that the more branches an isomer has, the more like a sphere it is and the lower its boiling temperature.

This means that the more branches an isomer has the less surface contact there is between the molecules. Less surface contact means weaker van der Waals forces between molecules and so less energy is needed to separate them – this results in a lower boiling temperature.

## The effect of functional groups on boiling temperatures

When comparing boiling temperatures of molecules with different functional groups the same arguments as mentioned for increasing chain length and branching apply and we also need to consider any other possible intermolecular forces. The strongest intermolecular forces are hydrgen bonds, therefore molecules that form hydrogen bonds have a higher boiling temperature than those of a similar size that cannot hydrogen bond.

The table below compares the boiling temperatures of compounds that have similar realtive molecular masses and almost the same number of electrons.

| Compound | Boiling temp / °C |
|---|---|
| Butane | −0.5 |
| Propan-1-ol | 97 |
| Ethanoic acid | 118 |

In butane, the only intermolecular forces are van der Waals induced dipole-induced dipole forces. In propan-1-ol, hydrogen bonding occurs between the –OH groups, so more energy is needed to break these bonds. In ethanoic acid, hydrogen bonding can occur in two places, so even more energy is needed to break these forces.

## Solubillity

The most significant intermolecular forces between water molecules are hydrogen bonds. Covalent compounds that can replace these bonds by forming new hydrogen bonds with water will dissolve.

Therefore hydrocarbons, like butane, cannot dissolve in water because they only form van der Waals interactions between molecules and are not able to form significant attractions with the polar water molecules.

However, alcohols and carboxylic acids have –OH groups that can form hydrogen bonds with water. Smaller alcohols and carboxylic acids are, therefore, soluble in water but solubility decreases as the chain length increases. As the size of the hydrocarbon part of the molecule increases it exerts such a large hydrophobic effect that the compound is no longer soluble.

# Test yourself

1. A student is analysing an unknown organic chemical X.

   (a) X contains carbon, hydrogen and oxygen only. It contains 55.0% carbon and 9.10% hydrogen by mass. Calculate the empirical formula of X. [3]

   (b) Given that the $M_r$ of compound X is 88, calculate the molecular formula for this compound. [1]

   (c) Fizzing occurs when X is mixed with sodium carbonate solution. State the name of the functional group that causes this observation. [1]

   (d) Name two isomers with the same functional group and molecular formula as X. [2]

   (e) Which of these two isomers would have the higher boiling point? Explain your answer. [2]

2. Propane-1,2,3-triol and 1,2,3-trichloropropane are both widely used in the chemical industry as solvents.

   (a) Draw the skeletal structure of propane-1,2,3-triol. [1]

   (b) What is the molecular formula of propane-1,2,3-triol? [1]

   (c) State the functional group present in propane-1,2,3-triol. [1]

   (d) Draw the displayed formula of 1,2,3-trichloropropane. [1]

   (e) To which homologous series does 1,2,3-trichloropropane belong? [1]

   (f) Which of the two compounds would have the higher boiling point? Explain your answer. [3]

3. Compound Z has the formula $CH_3CHBrCH_2Cl$. It has an interesting structure which lends itself to use in analytical chemistry.

   (a) What is the IUPAC name of compound Z? [1]

   (b) Draw the displayed formula and the skeletal formula of Z. [2]

   (c) What is the empirical formula? [1]

   (d) Identify the homologous series to which Z belongs. [1]

   (e) Name the four structural isomers of Z. [4]

# Hydrocarbons

Hydrocarbons are the naturally occurring raw materials for the production of many organic compounds. Petroleum, which contains hydrocarbons, can be fractionally distilled to produce fuels suitable for use in a wide variety of heating and other energy-dependent scenarios. It is also the starting point for the production of plastics and other synthetic materials.

In a world where consideration of the environmental issues associated with energy production is increasingly significant, it is necessary to thoroughly understand the nature of the reactions involved.

## Topic contents

You should be able to demonstrate and apply your knowledge and understanding of:

- Combustion reactions of alkanes, and benefits and drawbacks relating to the use of fossil fuels, including formation of carbon dioxide, acidic gases and carbon monoxide.
- C–C and C–H bonds in alkanes as σ-bonds.
- Mechanism of radical substitution, such as photochlorination of alkanes.
- Difference in reactivity between alkanes and alkenes in terms of the C=C bond as a region of high electron density.
- C=C bond in ethene and other alkenes as comprising π-bond and σ-bond.
- E-Z isomerism in terms of restricted rotation about a carbon to carbon double bond.
- Mechanism of electrophilic addition, such as in the addition of bromine to ethene, as a characteristic reaction of alkenes.
- Bromine/bromine water and potassium manganate(VII) tests for alkenes.
- Orientation of the normal addition of HBr to propene in terms of relative stabilities of the possible carbocations involved.
- Conditions required for the catalytic hydrogenation of ethene and the relevance of this reaction.
- Nature of addition polymerisation and the economic importance of the polymers of alkenes and substituted alkenes.
- Descriptions of species as electrophiles and radicals and bond fission as homolytic or heterolytic.

## Maths skills ≫

- Understand the symmetry of 2D and 3D shapes.
- Change the subject of an equation.
- Use ratios, fractions and percentages to determine empirical and molecular formulae.
- Use standard and ordinary form, decimal places and significant figures (this is part of section 2.4 in the specification, but is covered in this chapter).

# Fossil fuels

## Use of fossil fuels

Much of the world's industrial and domestic energy needs are met by the use of **fossil fuels**. These include natural gas, petroleum and coal. Although progress is being made in developing other sources, it seems we will be dependent, to an appreciable extent, on fossil fuels for the foreseeable future.

### Advantages of the use of fossil fuels

1 They are available in a variety of forms so that the type of fuel can be matched with its use. Coal and natural gas, for example, are widely used in power stations. Petroleum can be separated into fractions whose properties can be varied according to the purpose to which the fraction is to be put.

2 They are available at all times. Some green sources such as solar and wind have limited availability.

3 They are currently widely available and current infrastructure is generally set up around the use of fossil fuels.

### Disadvantages of the use of fossil fuels

1 They are, in practical terms, **non-renewable**. Theoretically, since animals and plants still die, they can be renewed. However, they are considered non-renewable since the formation of fossil fuels takes millions of years and reserves are being used faster than new ones are formed. This means that resources are running out.

2 Combustion of fossil fuels produces carbon dioxide. This is a **greenhouse gas** that acts by absorbing infrared radiation from the Earth's surface and then emitting it in all directions. Some of this radiation goes back towards the Earth's surface and therefore the surface temperature rises. This climate change has serious environmental consequences including rising sea levels and changes to crop suitability.

▲ Emissions of carbon dioxide from a factory and from a forest fire

> ### ▶▶ Key terms
>
> A **fossil fuel** is one that is derived from organisms that lived long ago.
>
> **Non-renewable resources** are those that cannot be reformed in a reasonable timescale.
>
> A **greenhouse gas** is one that causes an increase in the Earth's temperature.

> ### ▲ Stretch & challenge ▼
>
> Whilst carbon dioxide is usually quoted as the 'villain' in global warming, methane has 34 times the effect on temperature. The largest greenhouse effect is actually due to the presence, in the atmosphere, of water vapour.

> ### ◀◀ Exam tip
>
> Do not describe ozone depletion when asked about global warming. Ozone depletion is caused by CFCs.

## Knowledge check

**1**

Name two greenhouse gases present in the atmosphere.

## Knowledge check

**2**

Complete the sentence.

The presence of ............. acid and ............. acid in rain water ............. the pH.

## Knowledge check

**3**

Complete the equation to show what happens when acid rain runs over a marble statue.

$$CaCO_3(s) + \_\_\_ HNO_3(aq) \longrightarrow$$

............. + ............. + .............

## Study point

Many different equations can be used to show the formation of nitric acid in acid rain. These could include:

$$N_2(g) + O_2(g) \longrightarrow 2NO(g);$$

$$2NO(g) + O_2(g) \longrightarrow 2NO_2(g);$$

$$2NO_2(g) + H_2O(l) \longrightarrow HNO_3(aq) + HNO_2(aq).$$

You do not need to be able to quote a specific set of reactions in an exam.

## Study point

If incomplete combustion occurs, water is still always formed. It is the carbon part of the fuel that reacts with less oxygen so that carbon monoxide or even carbon is formed.

▲ The effects of acid rain

**3** Some of the products of combustion of fossil fuels contribute to the formation of acid rain. Many fossil fuels contain sulfur. On combustion, this produces sulfur dioxide. This reacts with water to make sulfuric(IV) (sulfurous) acid.

$$H_2O(l) + SO_2(g) \longrightarrow H_2SO_3(aq)$$

This is then oxidised to make sulfuric(VI) acid.

$$H_2SO_3(aq) + \tfrac{1}{2}O_2(g) \longrightarrow H_2SO_4(aq)$$

At the high temperatures of an internal combustion engine, atmospheric nitrogen and oxygen react to form oxides of nitrogen. These react with water to form nitric acid, $HNO_3$.

Sulfuric and nitric acids are present in acid rain. This can cause serious damage to buildings especially those that contain calcium carbonate.

The presence of sulfur dioxide and oxides of nitrogen is a health problem for people with breathing difficulties. During the Beijing Olympics, for example, road traffic was limited to avoid too much air pollution.

**4** Carbon monoxide formation. When fossil fuels are burned in an adequate supply of oxygen, complete combustion occurs and carbon dioxide and water are formed. However, if there is a shortage of oxygen, incomplete combustion occurs and carbon monoxide is formed. Carbon monoxide is toxic since it combines with haemoglobin in the blood so this is then not available to carry oxygen around the body.

The need to have adequate oxygen supplies for combustion means that air vents are fitted near boilers and instructions are attached to heaters that they should not be covered.

Incomplete combustion is also less efficient than complete combustion so that, for example, a car would achieve fewer miles per litre of fuel.

# Alkanes

## The homologous series of alkanes

1  The general formula of the homologous series is $C_nH_{2n+2}$.

2  Each member of the series differs from its neighbour by $CH_2$.

3  Alkanes all have similar chemical reactions as they are all saturated hydrocarbons.

4  The physical properties vary as the relative molecular mass increases. Small alkanes are gases at room temperature (e.g. natural gas is predominantly methane), larger ones are liquids (e.g. petroleum contains approximately 8 carbon atoms per molecule) and even larger ones are solids (e.g. wax candles).

## Reactions of alkanes

Since the electronegativities of carbon and hydrogen are similar, alkanes are non-polar. They also have no multiple bonds, only single (sigma) bonds, and so are generally unreactive. The σ-**bonds** are 'normal' covalent bonds. This includes bonds that involve s-s orbital overlap, end-on overlap of a p-orbital with an s-orbital and end-on overlap of two p-orbitals. Alkanes do, however, take part in two important reactions.

### 1 Combustion

Alkanes burn by reaction with oxygen. This reaction is exothermic so that alkanes are used as fuels.

If sufficient oxygen is present **complete combustion** occurs and carbon dioxide and water are formed.

Example using ethane:

$$C_2H_6(g) + 3\tfrac{1}{2}O_2(g) \longrightarrow 2CO_2(g) + 3H_2O(l)$$

If insufficient oxygen is present **incomplete combustion** occurs and carbon monoxide is formed.

Example using ethane:

$$C_2H_6(g) + 2\tfrac{1}{2}O_2(g) \longrightarrow 2CO(g) + 3H_2O(l)$$

Combustion reactions are often used in calculations.

### Practical check

The experimental procedure for the combustion of flammable liquids was covered in Section 2.1. This procedure could be used to determine the enthalpy of combustion for a liquid alkane such as octane.

You should be able to:

- Draw the apparatus required.
- Describe the method to obtain the data required for the calculations.
- Complete the calculations $q = mc\Delta T$ and $\Delta H = -q / n$.
- Suggest reasons why the result will not be accurate and give improvements to the method to make the result more accurate.

**Knowledge check**  4

Using knowledge of structures and bonding, explain why octane is a liquid at room temperature but methane is a gas.

**‹ Link ›**

Link to Section 2.4

**›› Key terms**

A σ **bond** is made by end to end overlap of s or p-orbitals.

**Complete combustion** is combustion that occurs with excess oxygen.

**Incomplete combustion** is combustion that occurs with insufficient oxygen.

**Knowledge check** 5

Why do cars that are incorrectly adjusted give out black smoke from their exhaust pipes?

**›› Study point**

Since alkanes are non-polar they cannot form hydrogen bonds with water. They are therefore insoluble in water. This can be seen when oil floats on top of water.

**◄ Stretch & challenge**

Although you might be asked to write or complete an equation showing incomplete combustion, really no wholly valid equation can be written. This is because, whenever carbon monoxide is formed, some carbon dioxide will also be formed. It is not possible to predict the ratio of each.

**Knowledge check**  6

Write the equation for the incomplete combustion of butane to form carbon monoxide and water.

**‹ Link ›**

Section 1.3 page 39

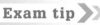

**7 Knowledge check**

Write the equation for the complete combustion of octane.

**Exam tip ≫**

Remember to make use of the data book in the exam. Many of these values are listed there plus there is useful information about how to convert from one unit to another.

**‹ Link ›**

Section 2.4 page 134

**Exam tip ≫**

This is not the molecular formula. To work out the molecular formula we would need to know the $M_r$ of the compound.

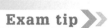

### Example 1 – to calculate the volume of gas produced

1.00g of methane is burned in excess oxygen, at atmospheric pressure and a temperature of 100°C. Calculate the volume of carbon dioxide produced and give your answer in cm³.

$$CH_4(g) + 2O_2(g) \longrightarrow CO_2(g) + 2H_2O(g)$$

This requires the use of the ideal gas equation because the temperature given is not 273K or 298K (in which case the molar gas volume could be used).

$PV = nRT$     rearrange     therefore $V = nRT / P$

$n$ = number of moles of $CH_4$ = number of moles of $CO_2$

    = $1 / 16.04 = 6.23 \times 10^{-2}$

$R$ = $8.31 \text{ J mol}^{-1} \text{ K}^{-1}$

$T$ = 373K

$P$ = $1.01 \times 10^5$ Pa

$$V = (6.23 \times 10^{-2}) \times 8.31 \times 373) / (1.01 \times 10^5)$$
$$= 1.91 \times 10^{-3} \text{ m}^3$$

To convert to cm³:

$$V = 1.91 \times 10^{-3} \times 10^6 = 1910 \text{ cm}^3$$

### Example 2 – to calculate the empirical formula of an unknown alkane

10.0g of an unknown alkane are burned in excess oxygen. The amounts of carbon dioxide and water produced are recorded. We can use this data to work out the empirical formula by the following method:

$$C_xH_y + \text{excess } O_2 \longrightarrow xCO_2 + \frac{y}{2}H_2O$$
$$\phantom{C_xH_y + \text{excess } O_2 \longrightarrow xCO_2 + } 2.00 \text{ g} \quad 1.23 \text{ g}$$

$$x = \text{mass} \frac{CO_2}{44} = 0.0455$$

$$\frac{y}{2} = (\text{mass} \frac{H_2O}{18.02})$$

$$y = (\text{mass} \frac{H_2O}{18.02}) \times 2 = 0.137$$

To find the simplest integer ratio:

| x | y |
|---|---|
| 0.0455 | 0.137 |
| 0.0455 / 0.0455 = 1 | 0.137 / 0.0455 = 3.01 |
| 1 | 3 |

Empirical formula = $CH_3$

## 2 Halogenation

**Halogenation** is the reaction between an organic compound and any halogen (member of Group 7).

Alkanes do not react with halogens in the dark. However, if exposed to UV light (often in the form of sunlight) they will react. The reaction takes place in three stages. The example below shows the reaction between methane and chlorine and is known as the photochlorination of methane.

### Stage 1 Initiation

**Initiation** starts the reaction. Ultraviolet light has sufficient energy to break the Cl — Cl bond homolytically. **Homolytic bond fission** occurs when each of the bonded atoms receives one of the bond electrons so that **radicals** are formed. Radicals are species that contain an unpaired electron. Radicals are highly reactive and they react indiscriminately to gain another electron to form an electron pair. In a radical only the unpaired electron is generally shown. Be careful with the terms *unpaired electron* and *lone pair of electrons*. Make sure you know what each means.

$$Cl_2 \longrightarrow 2Cl^\bullet$$

### Stage 2 Propagation

Radicals are very reactive and take part in a series of **propagation** reactions.

$$Cl^\bullet + CH_4 \longrightarrow CH_3^\bullet + HCl$$
$$CH_3^\bullet + Cl_2 \longrightarrow CH_3Cl + Cl^\bullet$$

A propagation reaction uses a radical as a reactant and then forms a radical as a product. This means that the reaction continues. It is therefore called a **chain reaction**.

### Stage 3 Termination

The propagation steps continue until two radicals meet. The reaction then stops – this is the **termination** stage.

$$Cl^\bullet + CH_3^\bullet \longrightarrow CH_3Cl$$

The stages initiation, propagation and termination are the three steps in the **reaction mechanism** and the overall mechanism is called radical substitution.

Exam questions may ask for the mechanism but be careful since sometimes a question asks for the overall equation. That is not the same as the mechanism. If, for example, a question asked for the equation for the reaction of ethane with bromine, the expected answer is just:

$$C_2H_6 + Br_2 \longrightarrow C_2H_5Br + HBr$$

Radical reactions are hard to control and further **substitution reactions** occur. This means that a mixture of products will be formed, which makes this an unsatisfactory method for making a high yield of a specific halogenoalkane. However, polysubstitution can be largely avoided if the amount of halogen used is very limited.

▶▶ **Key terms**

**Halogenation** is a reaction with any halogen.

**Initiation** is the reaction that starts the process. A molecule is turned into two radicals.

**Homolytic bond fission** is when a bond is broken and each of the bonded atoms receives one of the bond electrons.

**Radical** is a species with an unpaired electron.

**Propagation** is the reaction by which the process continues/grows. A molecule reacts with a radical to make a new molecule and a new radical.

A **chain reaction** is one that involves a series of steps and, once started, continues.

**Termination** is the reaction that ends the process. Two radicals react to make a molecule.

A **reaction mechanism** shows the stages by which a reaction proceeds.

A **substitution reaction** is one in which one atom/group is replaced by another atom/group.

**Knowledge check** 8 ◀

Complete the following:

Alkanes are generally unreactive because .............. Radicals are very reactive because .............

**Knowledge check** 9 ◀

Write the equation for the formation of tetrachloromethane from methane.

# Alkenes

## Structure of alkenes

Alkenes are an homologous series of unsaturated hydrocarbons. This means that they contain a carbon to carbon double bond and have the general formula $C_nH_{2n}$. Due to the presence of the double bond, alkenes are much more reactive than alkanes.

Alkenes are formed when petroleum is cracked and, although they will undergo combustion reactions, they are not generally used as fuels. They are, however, very important as the starting material for a variety of organic synthesis reactions including polymerisation.

### Ethene

In ethene, each carbon has the electronic structure $1s^2 2s^2 2p^2$, i.e. each has four electrons available for bonding. Three 'normal' covalent bonds are formed using three of these electrons. These bonds are called σ-bonds and they involve both s electrons and one of the p electrons in each carbon. This means that there are three bond pairs of electrons around each carbon, and VSEPR theory states that the shape around each will be trigonal planar with a bond angle of 120°. One p-orbital electron on each carbon is not used to form these bonds,

**10 Knowledge check**

Draw the displayed formula of (Z)-pent-2-ene..

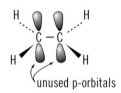

unused p-orbitals

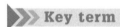

### Key term

A π-**bond** is one formed by the sideways overlap of p-orbitals.

The p-orbitals overlap **sideways** to produce a π-**bond** – an area of high electron density above and below the plane of the molecule.

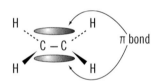

π bond

The restricted rotation about double bonds is due to the presence of the π-bond. Too much energy is needed to break the bond to allow free rotation. This means that many alkenes exist as *E-Z* isomers. See below.

### Other alkenes

Other alkenes also have bond angles of approximately 120° around the carbon atoms attached to the double bond but bond angles of approximately 109.5° about the carbon atoms in the saturated part of the chain.

**Link**

Link VSEPR page 55

For example, in propene:

# *E-Z* isomerism

Single bonds in alkanes allow the atoms/groups at either end to rotate freely but the double bond in alkenes means that rotation is restricted. This is due to the π bond.

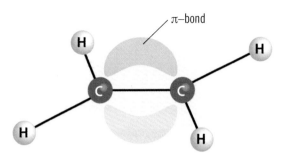

Since the double bond in alkenes and substituted alkenes restricts rotation, compounds such as 1,2-dibromoethene can exist in two different forms, i.e. two different isomers. These isomers are called **E-Z** isomers. *E-Z* isomers are only possible if each end of the C=C has two different groups bonded to it. The examples below show **E-Z isomerism** in 1,2-dibromoethene as well as the structure of 1,1-dibromoethene which does not show *E-Z*–isomerism.

(*Z*)-1,2-dibromoethene     (*E*)-1,2-dibromoethene     1,1-dibromoethene

▲ Isomers of dibromoethene

## Naming *E-Z* isomers

To decide which isomer is which you must look at the atom directly attached to each carbon in the double bond. Look at each carbon separately and the atom with the higher atomic number takes priority.

In 1,2-dibromoethene C 1 has a bromine and a hydrogen attached. Bromine has a higher atomic number and therefore takes priority.

C 2 also has a bromine and a hydrogen attached. Bromine has a higher atomic number and therefore again takes priority.

If both groups with higher priority are on the same side of the double bond the isomer is *Z*. This means that

is (*Z*)-1,2-dibromoethene.

> **Key term**
>
> **E-Z isomerism** is isomerism that occurs in alkenes (and substituted alkenes) due to restricted rotation about the double bond.

> **Study point**
>
> *Z* comes from *zusammen* (= together) and *E* comes from *entgegen* (= opposite).

> **Study point**
>
> In older textbooks you will see these types of isomers called *cis-trans*. This has now been superseded by *E-Z* since this has a wider scope of application.

> **Exam tip**
>
> An easy way to remember which isomer is which is to remember that in the *Z*-isomer, the groups are on 'ze zame zide', but in the *E*-isomer, they are 'enemies'.

If the groups with higher priority attached to the two carbons are on opposite sides of the double bond, it is the *E* isomer.

$$Br_{\diagdown}C=C_{\diagup}^{H}{\diagdown}Br$$

This is (*E*)-1,2-dibromoethene.

The naming system can apply to compounds containing double bonds with different groups attached.

$$Br_{\diagdown}C=C_{\diagup}^{Cl}{\diagdown}H$$

**Exam tip »**

Do not say that there is **no** rotation about the double bond. Rotation is possible but the large input of energy needed means that it is severely restricted.

Looking at C 1, bromine has a higher atomic number than hydrogen and therefore takes priority. In C 2, chlorine has a higher atomic number than hydrogen and takes priority. Both atoms with higher priority are on the same side of the double bond so that this is (*Z*)-1-bromo-2-chloroethene. Remember that bromo comes before chloro following the naming rules discussed on page 131.

The naming system can be extended to more complex molecules by looking further along the substituent chains.

Consider

$$CH_3{\diagdown}C=C_{\diagup}^{CH_2CH_3}{\diagdown}CH_3$$
$$CH_3CH_2^{\diagup}$$

Looking at C 3, it is attached directly to two carbon atoms – no priority can be assigned. You then have to look at these carbon atoms. In the $CH_3$ group the C is attached to H H H but in the $CH_3CH_2$ group this carbon is attached to C H H. This has a higher atomic number and so has priority. This isomer is therefore the *E* form because the two highest priority groups are on opposite sides and its name is (E)-3,4-dimethyl-hex-3-ene (following the naming rules on page 131).

## Properties of *E-Z* isomers

**11 Knowledge check**

Draw the *E* isomer of 1-bromo-2-chloroethene.

Melting temperature and boiling temperature depend on the strength of intermolecular forces. The strength of these forces is affected by how well the molecules pack together because this alters the area of contact between molecules.

In general the shape of *E* molecules means that they can fit together more closely. This means that they have stronger intermolecular forces and therefore higher melting temperatures. (*Z*)-butenedioic acid has a melting temperature of 130 °C whilst that of (*E*)-butenedioic acid is 286 °C.

**12 Knowledge check**

Does the structure below show the *E* or *Z* isomer?

# Reactions of alkenes

## Electrophilic addition reactions

Alkenes have a pair of electrons in a π orbital that makes them susceptible to attack by an electrophile, and so alkenes are far more reactive than alkanes. An **electrophile** is any species that can accept a lone pair of electrons. The mechanism of the reaction involves **heterolytic bond fission** leading overall to an **addition reaction**.

If, for example, hydrogen bromide is used, this has a permanent dipole and so attacks the π bond with the δ+ H being nearer to the negative area.

The mechanism in this case involves the use of 'curly arrows'. A curly arrow shows the movement of a **pair** of electrons.

However, if the alkene is unsymmetrical (the double bond is not in the middle of the molecule) then two halogenoalkane products are made (see next page for more details).

If a non-polar molecule, such as hydrogen or bromine, reacts with ethene a dipole is induced by the negative charge in the π bond.

$X_2$ could be $H_2$ or $Br_2$ to produce the overall equations

$$C_2H_4 + H_2 \longrightarrow C_2H_6$$
ethene                ethane

$$C_2H_4 + Br_2 \longrightarrow CH_2BrCH_2Br$$
ethene                1,2-dibromoethane

## Reaction of propene with hydrogen bromide

Consider the reaction between propene and hydrogen bromide. Two products are possible.

Propene                                      2-Bromopropane

Reaction 1

### Knowledge check  13

(a) Give the displayed formula of the product of the reaction of bromine with:
   (i) (Z)-1,2-dibromoethene
   (ii) (E)-1,2-dibromoethene
(b) How are the products of the reactions in (a) (i) and (ii) related to each other?

### Key terms

An **electrophile** is an electron-deficient species that can accept a lone pair of electrons.

**Heterolytic bond fission** is when a bond is broken and one of the bonded atoms receives both electrons from the covalent bond. Ions are formed.

An **addition reaction** is a reaction in which reagents combine to give only one product.

### Stretch & challenge

A 'curly arrow' with a full arrow head shows the movement of a pair of electrons. A 'curly arrow' with half an arrow head can be used to show the movement of a single electron.

$$Cl-Cl \longrightarrow Cl^{\bullet} + {}^{\bullet}Cl$$

### Study point

When drawing a mechanism you must be careful to show exactly where the 'curly arrow' starts and finishes. Arrows generally start from a π bond or another lone pair.

### Study point

Be careful that you use + or δ+ correctly. A positive charge,+, forms when a neutral species loses an electron; δ+ forms as part of a dipole.

**14** Knowledge check

Complete the equation:

$(CH_3)_2C = CH_2 + HBr \longrightarrow$

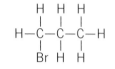

Propene           1-Bromopropane     Reaction 2

**Study point**

Delocalisation (the spread of a charge across several atoms) stabilises the ion. You will see this again when you consider aromatic compounds.

In practice the major product is 2-bromopropane as far more of it is formed than the 1-bromopropane. This can be explained by comparing the two carbocation intermediates formed in reactions 1 and 2.

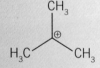 

Reaction 1   2° carbocation         Reaction 2   1° carbocation

**Key term**

A **carbocation** is an ion with a positively charged carbon atom.

The positive ion (the **carbocation**) is more stable in reaction 1 than in reaction 2. This is because alkyl groups tend to release electrons so that they become $\delta^+$. This spreads or delocalises the positive charge and this stabilises the ion. This means that secondary (2°) carbocations are more stable than primary (1°) carbocations.

**Study point**

In a test involving a change in colour you should include the original colour as well as what is seen in the test.

**Exam tip**

Questions in exams may ask you to draw the mechanism to show the formation of the major product and ask you to explain why this structure is formed in a greater amount than the alternative structure. Make sure you check for the two possible carbocation intermediates. Tertiary carbocations are more stable than secondary carbocations, which are more stable than primary carbocations.

Draw the mechanism which involves the most stable carbocation intermediate, and when explaining the formation of the major product remember to say that it is 'formed from' the more stable carbocation intermediate (tertiary or secondary as applicable). A common mistake is to say that the product is more stable rather than the carbocation intermediate is more stable.

tertiary carbocation    secondary carbocation    primary carbocation

▲ Types of carbocation

There are three types of carbocation and it is the number of alkyl groups which determines the type of carbocation. If there are three alkyl groups attached to the C⁺ then it is a tertiary carbocation, and so on.

## Uses of these reactions
### (i) Testing for alkenes

The reaction with bromine is used as the test for an alkene. Bromine is brown and therefore the presence of an alkene is shown by the decoloration of brown bromine water. Sometimes bromine water is used instead of liquid bromine in this test. This is because liquid bromine is very corrosive.

**15** Knowledge check

Suggest two reasons why bromine water, rather than chlorine, is used as the test for an alkene.

Another way of testing for alkenes is the decoloration of purple acidified potassium manganate(VII).

## (ii) Production of butter substitutes

The reaction with hydrogen can be called hydrogenation. It is catalysed by transition metals including platinum and palladium but nickel is most commonly used. The reaction is commercially important because liquid vegetable oils contain many double bonds (they are polyunsaturated) and these can be made saturated by adding hydrogen. This hardens the oil to make solid edible fats (margarines/butter substitutes).

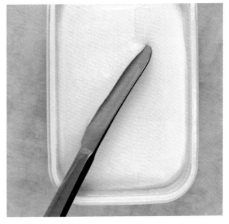

# Polymerisation

**Polymerisation** is the joining of a very large number of **monomer** molecules to make a large polymer molecule. Alkenes, and substituted alkenes, undergo addition polymerisation. In this type of reaction, the double bond is used to join the monomers and only one product forms.

When ethene is polymerised poly(ethene) is formed – this is commonly called polythene. In the equation below, showing this polymerisation, *n* is used to show a very large number.

$$n \; \begin{array}{c} H \\ | \\ C \\ | \\ H \end{array}=\begin{array}{c} H \\ | \\ C \\ | \\ H \end{array} \longrightarrow \left[ \begin{array}{cc} H & H \\ | & | \\ -C & -C- \\ | & | \\ H & H \end{array} \right]_n$$

The name of a polymer is formed by adding 'poly' to the name of the monomer. So although polythene has the suffix -ene, it doesn't have any double bonds along the carbon chain.

Poly(ethene) was discovered by accident when ethene and small amounts of other compounds were subjected to very high pressures and a white waxy material was formed. Initially it was difficult to control production and therefore the polymer had lots of side chains which prevented close packing, giving the polymer low density. With few points where the chains came close, there were limited Van der Waals forces that gave low melting points. These properties were not very useful in industrial terms. Once catalysts were found that enabled the production of straight chain poly(ethene), this gave an unreactive flexible solid with higher density and a higher melting point, and therefore with far more uses, such as plastic bags.

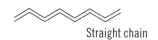

Straight chain

Branched chain

The differences between straight and branched chains when ethene was polymerised were an early example of how changing the structure of the polymer can change its physical properties to make it suitable for a particular use. This can also be achieved by using substituted alkenes as the monomers.

## Polymers of substituted alkenes

Since the monomers are joined using the double bond, polymers of substituted alkenes can be produced.

One method of finding the structure of the polymer formed from a particular monomer is to draw the monomer several times with the double bond in one molecule **next to** that in another molecule. The double bond breaks and is used to join the monomers together. This is shown by the arrows in the diagram.

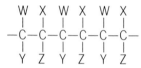

This gives the polymer as

W X W X W X
| | | | | |
—C—C—C—C—C—C—
| | | | | |
Y Z Y Z Y Z

The **repeat unit** of this polymer is

W X
| |
—C—C—
| |
Y Z

Economically important polymers include:

### (i) Poly(propene)

Monomer propene ⟶ Polymer poly(propene)

Poly(propene) is rigid and used in food containers and general kitchen equipment.

### (ii) Poly(chloroethene)

Monomer chloroethene ⟶ Polymer poly(chloroethene)

The old name for chloroethene was vinyl chloride, and poly(chloroethene) was called polyvinyl chloride (often abbreviated to PVC).

Poly(chloroethene) can be used in water pipes, in waterproof clothing or as the insulating covering for electrical cable.

**Knowledge check** 17

Draw 3 repeat units for the polymer you have named in KC 16. (The nitro group is $NO_2$.)

**Knowledge check** 18

What is the empirical formula of poly(ethene)?

## (iii) Poly(phenylethene)

Monomer
phenylethene

Polymer
poly(phenylethene)

**Study point**

When drawing a section of a polymer you must show the — at both ends to show that the chain continues. If you are asked to show only one repeat unit then '$n$' is not required. However, if you are asked to represent the whole polymer, then use brackets and put '$n$' outside the brackets.

The old name for phenylethene was styrene, and poly(phenylethene) was called polystyrene. This polymer is hard and is used in many household items needing strength and rigidity. It can be made into an insulator and packing material by creating holes in the structure. This is called expanded polystyrene.

**Knowledge check** 19

A section of polymer is shown below. Draw the monomer that was used to form this polymer.

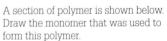

155

# Test yourself

1. Alkanes and alkenes are two types of hydrocarbons.

   (a) State the meaning of the term hydrocarbon. [1]

   (b) Give the general formula for:
       (i) alkanes [1]
       (ii) alkenes. [1]

   (c) State the key difference between alkanes and alkenes. [1]

   (d) Name the two types of bond present in alkenes. Explain how each type of bond is made. [3]

   (e) Use your answer to part (d) to explain the difference in reactivity between alkanes and alkenes. [2]

2. (a) Write an equation for the complete combustion of hexane. [1]

   (b) An unknown hydrocarbon was burned in excess oxygen and 6.1g of carbon dioxide was produced along with 3.0g of water. Calculate the empirical formula of this hydrocarbon. [3]

   (c) If insufficient oxygen is available for complete combustion, what gas is produced instead of carbon dioxide. Why is this gas a cause for concern? [2]

   (d) State two other issues related to the combustion of fossil fuels and explain why they are considered as disadvantages. [4]

3. (a) Name two straight chain isomers with the molecular formula $C_4H_8$. [1]

   (b) Which of these two isomers will display stereoisomerism? Explain your answer. [2]

   (c) When propene reacts with hydrogen bromide, two products are formed in different proportions.
       (i) Name the mechanism for this reaction. [1]
       (ii) Name the major product. [1]
       (iii) Draw the mechanism for the formation of the major product. [3]
       (iv) Explain why more of this product is formed compared to the alternative product.

# 2.6

# Halogenoalkanes

Halogenoalkanes are still used for a variety of industrial and household purposes but their use is now strictly controlled due to their toxicity and the environmental damage that they can cause. Understanding the chemistry of halogenoalkanes means that safer alternatives are now being developed.

As an homologous series, halogenoalkanes show reactions caused by nucleophilic substitution. These can be used for studies of the effect of changing the nature of the halogen or the structural isomer used on the rates of reactions.

You should be able to demonstrate and apply your knowledge and understanding of:

- Elimination reaction of halogenoalkanes forming alkenes, for example HBr eliminated from 1-bromopropane to form propene.
- Mechanism of nucleophilic substitution, such as the reaction between OH⁻(aq) and primary halogenoalkanes.
- Effect of bond polarity and bond enthalpy on the ease of substitution of halogenoalkanes.
- Hydrolysis/Ag⁺(aq) test for halogenoalkanes.
- Halogenoalkanes as solvents, anaesthetics and refrigerants, and tight regulation of their use due to toxicity or adverse environmental effects.
- Adverse environmental effects of CFCs and the relevance of the relative bond strengths of C–H, C–F and C–Cl in determining their impact in the upper atmosphere.
- How to carry out a reflux (for example, for nucleophilic substitution reaction of halogenoalkanes with hydroxide ions).
- Description of species as nucleophiles.

## Maths skills »

- Use standard and ordinary form, decimal places and significant figures.
- Use ratios, fractions and percentages.

## Knowledge check

**1**

What is the name of the following halogenoalkane?

$FCl_2CCH_2CHBrCH_3$

## Key terms

A **halogenoalkane** is an alkane in which one or more hydrogen atoms have been replaced by a halogen.

A **nucleophile** is a species with a lone pair of electrons that can be donated to an electron-deficient species.

**Reflux** is a process of continuous evaporation and condensation.

**Hydrolysis** is a reaction with water to produce a new product.

# Structure

**Halogenoalkanes** are an homologous series in which one or more of the hydrogen atoms have been replaced by a halogen. The series has the general formula $C_nH_{2n+1}X$ (where X is a halogen). The formula of a halogenoalkane is often shown as RX. Polysubstituted halogenoalkanes can also exist.

Halogenoalkanes contain a carbon to halogen bond. Since halogens are more electronegative than carbon, this bond is polar with a $\delta+$ carbon and a $\delta-$ halogen.

This dipole means that halogenoalkanes are susceptible to nucleophilic attack on the $\delta+$ carbon. This leads to substitution.

# Nucleophilic substitution

## Mechanism

A **nucleophile** has a lone pair of electrons that can be donated to an electron-deficient species.

Using $OH^-$ and 1-chloropropane as an example:

1-Chloropropane                    Propan-1-ol

## Preparation of alcohols

Nucleophilic substitution can be used to prepare alcohols from halogenoalkanes, e.g. butan-1-ol from 1-bromobutane. The nucleophile $OH^-$ is provided by aqueous sodium hydroxide.

$$CH_3CH_2CH_2CH_2Br + OH^- \longrightarrow CH_3CH_2CH_2CH_2OH + Br^-$$

The mechanism for the reaction between aqueous hydroxide ions and a halogenoalkane is classified as nucleophilic substitution. However, since the reaction can also be achieved using water (which contains a small amount of $OH^-$), it can also be classified as **hydrolysis**. Other nucleophiles such as $CN^-$ (cyanide) and ammonia will also react with halogenoalkanes. These reactions are studied in more detail at A2 level.

Since the reaction is rather slow it is necessary to heat the mixture of 1-bromobutane and aqueous sodium hydroxide for a significant time. If this were done in an open flask or beaker much of the liquids would evaporate and be lost and the yield of product would be very low. To avoid this problem the liquid mixture is **refluxed**. Refluxing allows prolonged heating of volatile chemicals without loss.

When heated, the liquids evaporate and vapour escapes from the flask. However, when the vapour reaches the condenser it condenses, and liquid forms. This liquid drips back to the reaction flask. This means that in this process of continuous evaporation and condensing the liquid can be boiled for as long as is needed to achieve a reaction without any loss of material.

## Exam tip

There are three mechanisms in Unit 2: radical substitution, electrophilic addition, and nucleophilic substitution. One will always be on the exam paper so learn them carefully.

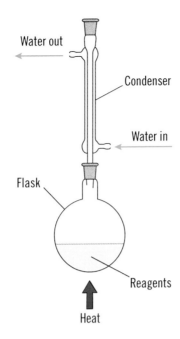

**Knowledge check** 2

Name the reagents that could be used to prepare 2-methylbutan-2-ol.

**Knowledge check** 3

State the conditions required for the reaction in KC 2.

**Knowledge check** 4

Write the equation for the reaction in KC 2.

When liquids are refluxed, the top of the condenser must be open. If it is not, when the apparatus is heated, the air will expand and blow the stopper off – usually violently!

### Practical check ≫

The refluxing of a halogenoalkane to make an alcohol is a **specified practical** (full details can be found in the Year 12 section of the Lab Book). Understanding how the preparation occurs requires understanding of the principles of both refluxing and distillation – you should be able to draw labelled diagrams for both techniques. More information on distillation can be found in Topic 2.7 of this book.

#### Example

When the alcohol is first distilled over, it is obtained as a mixture of water and butan-1-ol. As the alcohol has limited solubility in water, the mixture separates into two layers and the alcohol layer has a volume of 7.25 cm³. The density of butan-1-ol is 0.81 g/cm³.

1  Draw a labelled diagram of the layers of water and butan-1-ol in a conical flask.

2  Calculate the number of moles of butan-1-ol obtained.

3  If 15.1 g of 1-bromobutane was used at the start, calculate the yield of this preparation.

#### Answers:

1

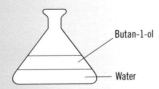

2  Number of moles of butan-1-ol = mass / $M_r$ (where mass = volume × density)

$$= (7.25 \times 0.81) / 74.1$$
$$= 7.93 \times 10^{-2}$$

3  There is a 1:1 ratio of halogenoalkane to alcohol.

Theoretical number of moles that could be produced = 15.1 / 136.99

$$= 1.10 \times 10^{-1}$$

Therefore, yield = $\dfrac{7.93 \times 10^{-2}}{1.10 \times 10^{-1}} \times 100 = 72.1\%$

# Effect of changing the halogen

A similar reaction occurs for all the halogens but the rate at which the reaction occurs depends on the nature of the halogen. Using chloro, bromo and iodo compounds there are two factors that have to be considered.

## (i) Electronegativity

Electronegativity decreases as the size of the halogen increases. This means that C–Cl is the most polar bond with this carbon atom being the most $\delta+$.

## (ii) Bond strength

The substitution reaction involves breaking the carbon to halogen bond. This bond is strongest for C–Cl and this therefore is the most difficult bond to break.

The effects in (i) and (ii) act in opposite directions and so it is difficult to predict which of the halogenoalkanes would be hydrolysed most rapidly. In fact (ii) dominates so that the order of rate of hydrolysis for corresponding halogenoalkanes is:

fastest             slowest

iodo > bromo > chloro

The rate of the reaction can be followed using the fact that after hydrolysis a halide ion is formed in solution. This can be detected by adding $Ag^+(aq)$, usually as aqueous silver nitrate, and timing how long the precipitate takes to form. The technique of colorimetry could be used to monitor the progress of the reaction. This is discussed in more detail in Section 2.2.

# Test for halogenoalkanes

Water can be used to hydrolyse the halogenoalkane but this is rather slow and therefore aqueous sodium hydroxide is often used and the reaction is heated. Before aqueous silver nitrate is added the excess sodium hydroxide must be neutralised by adding dilute nitric acid, as sodium hydroxide would interfere with the test. If dilute nitric acid is not added before the aqueous silver nitrate, silver hydroxide, as a brown precipitate, will be formed.

$$RX + NaOH(aq) \longrightarrow ROH + Na^+(aq) + X^-(aq)$$

The presence of the halide ion is then shown by the aqueous silver nitrate in the usual test.

$$X^-(aq) + Ag^+(aq) \longrightarrow AgX(s)$$

| Halogen in halogenoalkane | Addition of $Ag^+(aq)$ | Addition of $NH_3(aq)$ to precipitate formed with $Ag^+(aq)$ |
|---|---|---|
| chlorine | white precipitate | dissolves in dilute $NH_3(aq)$ |
| bromine | cream precipitate | dissolves in concentrated $NH_3(aq)$ |
| iodine | yellow precipitate | does not dissolve in $NH_3(aq)$ |

**Knowledge check** 5

Why is chlorine more electronegative than bromine?

**Knowledge check** 6

Why is the C–Cl bond stronger than the C–Br bond?

**Study point**

The test for a halogenoalkane is in two parts. First is the formation of the halide ion and this is then followed by testing for this ion.

**Link**

Test for halide ions page 71

**Knowledge check** 7

Write down, in order, the steps needed to test for the presence of bromine in an organic compound. Include the result expected.

**Knowledge check** 8

Write the ionic equation for the reaction that occurs to produce the yellow precipitate formed if iodine is present in an organic compound.

# Elimination reactions

An **elimination reaction** is one that involves the loss of a small molecule to produce a multiple bond.

As well as the substitution reactions seen above, halogenoalkanes can undergo elimination reactions. Whether substitution or elimination occurs depends upon the conditions used.

Using 1-bromopropane as an example, an elimination reaction is shown in the equation:

$$\begin{array}{c} \text{H} \;\; \text{H} \;\; \text{H} \\ | \;\;\; | \;\;\; | \\ \text{H--C--C--C--Br} \\ | \;\;\; | \;\;\; | \\ \text{H} \;\; \text{H} \;\; \text{H} \end{array} + \text{NaOH} \longrightarrow \begin{array}{c} \text{H} \;\; \text{H} \\ | \;\;\; | \\ \text{H--C--C=C} \\ | \;\;\; | \\ \text{H} \;\; \text{H} \end{array}\!\!{}^{\text{H}}_{\text{H}} + \text{NaBr} + \text{H}_2\text{O}$$

1-Bromopropane                           Propene

The alkali, such as sodium hydroxide, must be dissolved in ethanol without any water present to avoid the substitution reaction.

In order to eliminate a hydrogen halide, the halogen must be attached to a carbon which is next to a carbon that has a hydrogen attached. This means that structures like 1-bromo-dimethylpropane will not undergo elimination reactions. Look at the structure (right) to make sure you understand why.

$$\begin{array}{c} \text{CH}_3 \\ | \\ \text{CH}_3\text{--C--CH}_2\text{Br} \\ | \\ \text{CH}_3 \end{array}$$

If the halogenoalkane is unsymmetrical, more than one alkene can be formed by elimination reactions. An example is:

$$\begin{array}{c} \text{H} \;\; \text{H} \;\; \text{H} \;\; \text{H} \\ | \;\;\; | \;\;\; | \;\;\; | \\ \text{H--C--C--C--C--H} \\ | \;\;\; | \;\;\; | \;\;\; | \\ \text{H} \;\; \text{Br} \;\; \text{H} \;\; \text{H} \end{array}$$

and

$$\begin{array}{c} \text{H} \;\;\;\;\;\;\;\;\; \text{H} \;\; \text{H} \;\; \text{H} \\ \text{\textbackslash} \;\;\;\;\;\; | \;\;\; | \;\;\; | \\ \text{C=C--C--C--H} \\ / \;\;\;\;\;\;\; | \;\;\; | \\ \text{H} \;\;\;\;\;\;\;\;\; \text{H} \;\; \text{H} \end{array} + \text{HBr}$$

But-1-ene

$$\begin{array}{c} \text{H} \;\;\;\;\;\;\; \text{H} \;\; \text{H} \\ | \;\;\;\;\;\;\;\; | \;\;\; | \\ \text{H--C--C=C--C--H} \\ | \;\;\;\;\;\;\;\; | \\ \text{H} \;\; \text{H} \;\;\;\;\;\; \text{H} \end{array} + \text{HBr}$$

But-2-ene

You do not need to know which product would dominate.

## Summary

To convert halogenoalkanes to alcohols, aqueous sodium hydroxide and heat are used. This substitution reaction involves a nucleophile. However, if we want to make an alkene we heat the halogenoalkane with sodium hydroxide in pure ethanol. This time the mechanism is elimination and the hydroxide ion behaves as a base rather than a nucleophile.

**Knowledge check**  **9**

What is the product(s) of the reaction of 1-bromo-2-methylbutane with:

(a) Aqueous sodium hydroxide?

(b) Ethanolic sodium hydroxide?

**Knowledge check** **10**

What is the product(s) of the reaction of 2-bromo-2-methylbutane with:

(a) Aqueous sodium hydroxide?

(b) Ethanolic sodium hydroxide?

# Uses of halogenoalkanes

### 1 Solvents

Halogenoalkanes contain a polar section due to the presence of the carbon to halogen bond but they also contain a non-polar section due to the presence of the alkyl chain. This means that halogenoalkanes can mix with a variety of polar and non-polar organic substances and are therefore used as solvents. The non-flammability of halogenoalkanes is also an advantage enabling them to be used in dry cleaning clothes and other processes involving degreasing.

Chlorocompounds such as tetrachloromethane, $CCl_4$, and tetrachloroethene, $CCl_2 = CCl_2$, were widely used in the past although concerns later emerged about their toxicity. Chlorocompounds were preferred to other halogenoalkanes because they were cheaper.

### 2 Anaesthetics

Many halogenoalkanes can act as general anaesthetics. Trichloromethane, $CHCl_3$, (chloroform) was one of the earliest substances to be used and this revolutionised surgical procedures. Another famous 'knock out' drug used in the past was a Mickey Finn. This consists of 2,2,2-trichloroethane-1,1-diol, $CCl_3CH(OH)_2$, (chloral hydrate). Although these compounds are no longer commonly used, halothane, $CF_3CHBrCl$, is still used in anaesthesia.

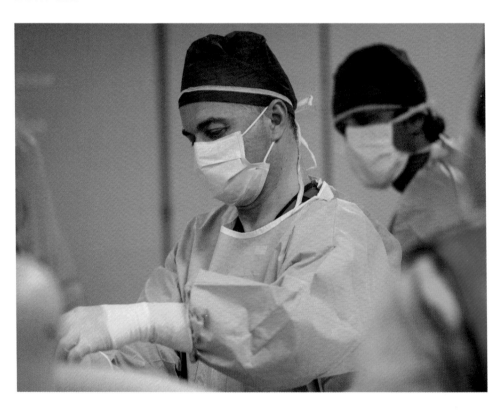

### 3 Refrigerants

Although small halogenoalkanes are gases at room temperature, the presence of permanent dipole-permanent dipole attractions means that their boiling temperatures are close to room temperature. They are therefore liquids that can easily be evaporated or gases that can easily be liquefied at room temperature. Chlorofluorocarbons, **CFCs**, were used as refrigerants since the heat needed to change the liquid to the gas is removed from the fridge, cooling the fridge contents. Non-flammability and non-toxicity, as well as an appropriate boiling temperature, are important properties in making these halogenoalkanes suitable for this use.

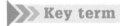

### Key term

CFCs are halogenoalkanes containing both chlorine and fluorine.

# Adverse environmental effects of CFCs

Nowadays the use of halogenoalkanes is limited by statutory regulations. This is because many of them, particularly polychloroalkanes, have been found to be toxic and others, such as the CFCs, cause damage to the Earth's **ozone layer**. The use of CFCs is thought to have caused holes in the ozone layer, particularly over the Arctic and Antarctic.

The ozone layer prevents harmful UV radiation reaching the Earth's surface. Therefore, depletion of the ozone layer results in penetration to the Earth of UV rays which could cause skin cancer. The damage process involves radical substitution reactions and, as seen with halogenation of alkanes, the mechanism includes a series of steps.

> **Key terms**

**Ozone layer** a layer surrounding the earth that contains $O_3$ molecules.

**HFCs** halogenoalkanes containing fluorine as the only halogen.

**Link**

Halogenation of alkanes on page 147

## 1 Initiation stage

In the upper atmosphere, UV radiation causes homolytic bond fission of the C–Cl bond in the CFC and produces chlorine radicals. This can be shown using the CFC dichlorodifluoromethane as an example:

$$CCl_2F_2 \longrightarrow Cl^{\bullet} + CClF_2^{\bullet}$$

## 2 Propagation stages

There are many possible propagation stages which include:

$$Cl^{\bullet} + O_3 \longrightarrow ClO^{\bullet} + O_2$$

$$ClO^{\bullet} + O_3 \longrightarrow Cl^{\bullet} + 2O_2$$

Since these form a chain reaction, the presence of a small number of chlorine radicals can cause the decrease of many ozone molecules.

> **Study point**

Do not try to remember particular propagation stages. Many alternatives are possible but they all convert $O_3$ to $O_2$ whilst regenerating the chlorine radical.

## Alternatives to CFCs

To avoid the formation of chlorine radicals, alternative compounds that do not contain chlorine are being increasingly used. These include alkanes (although these have the problem of inflammability) and **HFCs**. Since the only halogen in HFCs is fluorine, radicals are not formed when they are exposed to UV radiation.

The table shows the bond strengths for bond types commonly found in halogenoalkanes.

**Knowledge check**

Why is the C–F bond stronger than the C–Cl bond?

| Bond | Bond strength/ $kJ\,mol^{-1}$ |
|------|-------------------------------|
| C–H  | 413 |
| C–Cl | 328 |
| C–F  | 485 |

Comparing the bond strengths explains why UV radiation splits C–Cl bonds but not C–F bonds. Given that the majority of UV radiation which penetrates the Earth's atmosphere has wavelengths between 300 and 400 nm, we can calculate that there is insufficient energy to break the stronger C–F bonds but enough energy to break the C–Cl bonds.

# Test yourself

1. Halogenoalkanes undergo various reactions and can be converted into useful products.

   (a) Name the mechanism that occurs when halogenoalkanes react with sodium hydroxide in aqueous conditions. [1]

   (b) What is meant by the term nucleophile? [1]

   (c) What feature of halogenoalkanes makes them susceptible to attack by nucleophiles? Explain why this feature arises. [2]

   (d) Draw the mechanism for the reaction of 2-chloropropane with aqueous sodium hydroxide. [3]

   (e) Draw the skeletal formula of the product. [1]

   (f) Name a suitable halogenoalkane that would be required to make 2-methylbutan-2-ol. [1]

   (g) State the reaction conditions required to convert a halogenoalkane into an alkene and name this mechanism. [2]

   (h) Name the alkenes produced when 2-bromobutane is reacted with sodium hydroxide under these conditions. [2]

2. 1-Bromopropane can be converted into propan-1-ol by reacting with aqueous sodium hydroxide.

   (a) Draw the apparatus and explain how this method ensures that the reaction occurs as fully as possible. [5]

   (b) Name the technique that could be used to separate the product. Explain fully why this technique can be used to separate the mixture. [3]

   (c) Describe the chemical test, and give the result, that could be carried out to prove that hydrolysis has occurred. [3]

   (d) Would hydrolysis be faster or slower if 1-iodopropane was used instead? Explain your answer. [2]

3. Halogenoalkanes can be used as solvents, anaesthetics and refrigerants; however, their use is tightly controlled due to adverse environmental effects.

   (a) Describe why scientists have been concerned about the release of halogenoalkanes into the atmosphere. [4]

   (b) How have halogenoalkanes been modified to avoid this adverse environmental effect? Explain why these alternatives do not cause the same problem. [2]

   (c) Given that the bond enthalpy of C–H is 413 kJ mol$^{-1}$, explain fully why light with a wavelength of 400 nm will not break C–H bonds. [4]

# 2.7

# Alcohols and carboxylic acids

Everyone will be aware of some alcohols and carboxylic acids. To many people 'alcohol' refers to the ethanol that is present in alcoholic drinks but to a chemist the word refers to compounds with the OH functional group.

Carboxylic acids also contain OH but in this case it is within the COOH functional group and this gives the acids very different properties from the alcohols. Carboxylic acids are widely distributed in 'acidic' foodstuffs including vinegar, lemons and sour milk.

The reaction between an alcohol and a carboxylic acid gives an ester – esters are widely used as the active part of perfumes and flavourings.

You should be able to demonstrate and apply your knowledge and understanding of:

- Industrial preparation of ethanol from ethene.
- Preparation of ethanol by fermentation followed by distillation, and issues relating to the use of biofuels.
- Dehydration reactions of alcohols.
- Classification of alcohols as primary, secondary and tertiary.
- Oxidation of primary alcohols to aldehydes/carboxylic acids and secondary alcohols to ketones.
- Dichromate(VI) test for primary/secondary alcohols and sodium hydrogencarbonate test for carboxylic acids.
- Reactions of carboxylic acids with bases, carbonates and hydrogencarbonates forming salts.
- Esterification reaction that occurs when a carboxylic acid reacts with an alcohol.
- Separation by distillation.

## Maths skills ≫

- Use logarithmic functions on calculators.

# Alcohols

**Alcohols** are an homologous series in which one of the hydrogen atoms in the alkane has been replaced by –OH. This means that –OH is the functional group. Alcohols also exist that contain more than one –OH group but ethanol is the most widely used alcohol. Ethanol is what, in everyday language, is called alcohol and it is present in alcoholic drinks. The alkyl group can be shown as R so that an alcohol would be ROH.

# Industrial preparation of ethanol

Since ethanol is industrially and commercially important it is necessary to manufacture it on a large scale. This is achieved in two main processes.

## (a) Hydration of ethene

Ethene is obtained, on a large scale, by cracking hydrocarbons produced from petroleum. It reacts with steam to produce ethanol.

$$CH_2=CH_2(g) + H_2O(g) \rightleftharpoons CH_3CH_2OH(g) \qquad \Delta H = -45\,kJ\,mol^{-1}$$

The conditions generally used in this conversion are a temperature of about 300 °C, a pressure of about 60–70 atmospheres and a catalyst of phosphoric acid (coated onto an inert solid). These conditions can be explained using Le Chatelier's principle.

### Temperature

Since the forward reaction is exothermic, a high yield would be favoured by a low temperature. This, however, gives a slow rate of reaction: 300 °C is a compromise temperature.

### Pressure

In the reaction, two moles of gas react to produce one mole of gas. A high yield is therefore favoured by a high pressure. High pressure also increases the rate of reaction but if the pressure is too high, more powerful pumps and stronger pipes are needed and this increases costs; 60–70 atmospheres is therefore used.

### Catalyst

The catalyst does not affect the yield but it does increase the rate at which ethene and steam react to produce the equilibrium concentration of ethanol.

Under these conditions only about 5% of the ethene is converted and therefore the remaining ethene is recycled back to the reaction chamber.

## (b) Fermentation

**Fermentation** is the process by which sugars are converted into ethanol. The reaction is generally carried out by dissolving the sugar in water, adding yeast and leaving the mixture

**Key terms**

**Alcohol** is an homologous series containing —OH as the functional group.

**Fermentation** is an enzyme-catalysed reaction that converts sugars to ethanol.

**Knowledge check**

How could ethanol be removed from the equilibrium mixture produced when ethene reacts with steam?

**Link**

Link Le Chatelier's principle page 78

**Stretch & challenge**

The mechanism for the hydration of ethene involves electrophilic attack on the π bond by the δ+ H on the water. Try to draw this mechanism using curly arrows.

in a warm place. Yeast contains enzymes that catalyse the reaction. Using glucose as an example, the equation for the reaction is:

$$C_6H_{12}O_6 \longrightarrow 2C_2H_5OH + 2CO_2$$

Fermentation is the method by which alcoholic drinks are produced.

Carbon dioxide escapes as a gas but the ethanol has to be separated from the remaining liquid mixture. The boiling temperature of ethanol is 78 °C and therefore to separate it from an aqueous mixture, fractional distillation in needed.

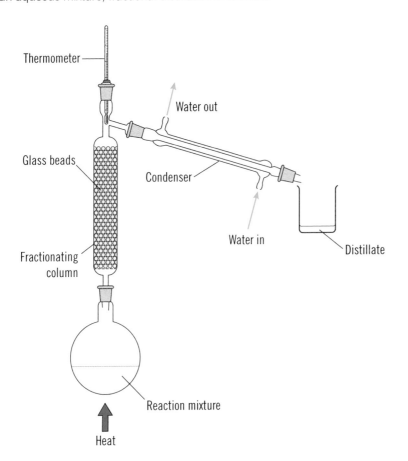

Thermometer

Water out

Glass beads

Condenser

Fractionating column

Water in

Distillate

Reaction mixture

Heat

**≪ Practical check**

Nothing apart from glucose is needed as a reagent in fermentation. Water is needed for the reaction to proceed but it is not part of the overall equation.

**Knowledge check 2◀**

Why is fractional distillation, rather than simple distillation, needed in this separation?

**≫ Key term**

**Biofuel** is a fuel that has been produced using a biological source.

# Biofuels

**Biofuels** are fuels that are produced from living organisms. The two main types currently in use are bioethanol and biodiesel. Bioethanol is obtained from sugars in plants by fermentation (as described above) and biodiesel is obtained from the oils and fats present in the seeds of some plants.

Since both bioethanol and biodiesel burn exothermically they can be used as fuels. If used in internal combustion engines they are generally blended with conventional fossil fuels.

# Issues relating to the use of biofuels

There are advantages and disadvantages in using biofuels compared with using fossil fuels.

## Advantages

### (i) Renewable

Fossil fuels are non-renewable and will eventually run out. Plants can be grown each year and biofuels can also be produced using waste material from animals.

### (ii) Greenhouse gases

Overall production of carbon dioxide is reduced. Although carbon dioxide is produced in exactly the same way by biofuels and fossil fuels when combusted, carbon dioxide is taken in by the plants when they are grown. Plants use carbon dioxide in photosynthesis to produce sugars.

$$6CO_2 + 6H_2O \longrightarrow C_6H_{12}O_6 + 6O_2$$

This means that, since carbon dioxide is produced in one process but removed in another, overall the use of biofuels can be considered to be **carbon neutral**. However, many other factors must be also be taken into account – see below.

### (iii) Economic and political security

Countries that are able to grow crops such as sugar cane, and create their own biofuel, do not need to use fossil fuels. This is especially relevant to those countries that do not have fossil fuels as a natural resource. Countries producing biofuels are less affected by changes in fossil fuel prices and changes in its availability.

## Disadvantages

### (i) Land use

Land that is used to produce plants for biofuels cannot be used to produce food. There is also pressure to destroy environmentally significant areas, such as forests, to create land for biofuel production.

### (ii) Use of resources

Growing crops suitable for biofuels needs large quantities of water and fertilisers. The use of water can strain local resources and the use of large quantities of fertilisers to grow the same crop year after year can cause water pollution.

### (iii) Carbon neutrality

When the fuel needed to build and run the factories for biofuel production, to transport raw materials and finished products, etc., is considered, it can be argued that the use of such fuels is not carbon neutral.

# Dehydration of alcohols

Many alcohols can be dehydrated to form alkenes. The equation for propan-1-ol is shown below:

$$CH_3CH_2CH_2OH \longrightarrow CH_3CH = CH_2 + H_2O$$
propan-1-ol            propene

Many different dehydrating agents are possible but concentrated sulfuric acid or heated aluminium oxide are most commonly used.

The reaction is elimination since a double bond is produced by the removal of OH from one carbon atom and of H from an adjacent carbon atom. The structure of the alcohol determines what alkene is made, so it is possible to produce more than one alkene from some alcohols. Remember, if there is no H atom on the adjacent carbon to the carbon atom with the OH group then elimination cannot occur.

**Key term**

A **Carbon neutral** process is where there is no net transfer of carbon dioxide to or from the atmosphere.

**Link**

Section 2.3

**Study point**

The dehydration of alcohols is basically the same reaction as the elimination of hydrogen halides from halogenoalkanes. Both reactions involve the formation of alkenes.

**3 Knowledge check**

Write the equation for the reaction that occurs when 2-methylpropan-1-ol is passed over heated aluminium oxide.

# Classification of alcohols

Alcohols are **classified** as being primary, 1°, secondary, 2°, or tertiary, 3°, according to the bonding of the – OH in the molecule.

**Primary alcohol**: the –OH is joined to a carbon that is itself joined to not more than one other carbon atom.

**Secondary alcohol**: the –OH is joined to a carbon that is itself joined to two other carbon atoms.

**Tertiary alcohol**: the –OH is joined to a carbon that is itself joined to three other carbon atoms.

## Examples of classification

*(a) Primary alcohols*

(i) Methanol                    (ii) Propan-1-ol

The OH is joined to a carbon that is itself joined to not more than one other carbon atom.

*(b) Secondary alcohols*

Propan-2-ol

The OH is joined to a carbon attached to two other carbons.

*(c) Tertiary alcohols*

2-methylbutan-2-ol

The OH is joined to a carbon attached to three other carbons.

**Knowledge check**

Draw all the structures of the elimination products formed from

(a) butan-2-ol

(b) dimethylpropan-1-ol

**Knowledge check**

Classify the following as 1°, 2° or 3° alcohols.

(a) Aminomethanol, $NH_2CH_2OH$

(b) 2-methylbutan-2-ol, $CH_3C(CH_3)(OH)CH_2CH_3$

(c) Butan-2-ol $CH_3CH(OH)CH_2CH_3$

(d) Butan-1-ol, $CH_3CH_2CH_2CH_2OH$

# Oxidation of alcohols

Acidified potassium dichromate(VI) can be used to oxidise many alcohols. Since dichromate(VI) will only behave satisfactorily as an oxidising agent in the presence of $H^+$, the oxidation is generally carried out by heating the alcohol with a mixture of aqueous potassium dichromate(VI) and sulfuric acid.

In equations showing oxidation of organic compounds the oxidising agent is usually shown as [O].

What happens under these conditions depends on whether the alcohol is primary, secondary or tertiary.

### Primary – using ethanol as an example

The reaction takes place in two stages.

*Stage 1*

$$H-\overset{\overset{\displaystyle H}{|}}{\underset{\underset{\displaystyle H}{|}}{C}}-\overset{\overset{\displaystyle H}{|}}{\underset{\underset{\displaystyle H}{|}}{C}}-O-H \quad + \quad [O] \quad \longrightarrow \quad H-\overset{\overset{\displaystyle H}{|}}{\underset{\underset{\displaystyle H}{|}}{C}}-C\!\!\begin{array}{c}\nearrow O\\ \searrow H\end{array} \quad + \quad H_2O$$

Ethanol                                        Ethanal

Two hydrogen atoms are lost – one from the alcohol OH and one from the adjacent carbon. This creates a carbon to oxygen double bond.

The product, with the functional group $-C\!\!\begin{array}{c}\nearrow O\\ \searrow H\end{array}$

is an **aldehyde**. In this case ethanal.

*Stage 2*

The aldehyde is oxidised further.

$$H-\overset{\overset{\displaystyle H}{|}}{\underset{\underset{\displaystyle H}{|}}{C}}-C\!\!\begin{array}{c}\nearrow O\\ \searrow H\end{array} \quad + \quad [O] \quad \longrightarrow \quad H-\overset{\overset{\displaystyle H}{|}}{\underset{\underset{\displaystyle H}{|}}{C}}-C\!\!\begin{array}{c}\nearrow O\\ \searrow OH\end{array}$$

Ethanal                                        Ethanoic acid

An oxygen is added to the aldehyde.

The product, with the functional group $-C\!\!\begin{array}{c}\nearrow O\\ \searrow O-H\end{array}$

is a **carboxylic acid**. In this case ethanoic acid.

## Secondary – using propan-2-ol as an example

The reaction has only one stage – this corresponds to stage 1 of the oxidation of primary alcohols.

Propan-2-ol                                                    Propanone

The product, with the functional group –C=O, is a **ketone**, in this case propanone.

## Tertiary – using 2-methylpropan-2-ol as an example

$$CH_3-\underset{\underset{CH_3}{|}}{\overset{\overset{\overset{\displaystyle H}{|}}{\displaystyle O}}{\overset{|}{C}}}-CH_3$$

2-Methylpropan-2-ol

Since there is no hydrogen on an adjacent carbon that can be lost, no reaction takes place.

## Summary of oxidation reactions

Primary alcohols $\longrightarrow$ aldehydes $\longrightarrow$ carboxylic acids

Secondary alcohols $\longrightarrow$ ketones

Tertiary alcohols are not oxidised.

## Use of acidified potassium dichromate(VI)

When it behaves as an oxidising agent, acidified potassium dichromate(VI) changes colour from orange to green. This can be used as a test for primary and secondary alcohols since they will give a positive test result but tertiary alcohols will not.

>> **Study point**

In the oxidation equation for stage 1 you must not forget to balance the equation by including the water produced.

>> **Key term**

**Ketone** is an homologous series containing a C=O group within a carbon chain.

**Knowledge check**

Write an equation for the oxidation of propan-1-ol to propanal.

**Knowledge check** 7

Write an equation for the oxidation of butan-1-ol to butanoic acid.

**Knowledge check**

What apparatus would you use if you wanted to prepare a carboxylic acid from an alcohol? Explain your answer.

# Carboxylic acids

Carboxylic acids are an homologous series that contain the functional group $-C\overset{\displaystyle O}{\underset{\displaystyle O-H}{\phantom{|}}}$.

This is often written as —COOH or —CO$_2$H. Many commonly used substances contain carboxylic acids and the acids often have names that refer to these sources. The names used in scientific situations are, of course, based on the rules for naming organic compounds that are given in Section 2.4. Some examples of carboxylic acids with both their chemical and common names are given in the table.

| Chemical name | Commonly used name | Formula | Present in | Reason for name |
|---|---|---|---|---|
| Methanoic acid | Formic acid | HCOOH | Venom from ant bites | Latin for ant is formica |
| Ethanoic acid | Acetic acid | CH$_3$COOH | Vinegar | |
| 2-hydroxypropane-1,2,3-tricarboxylic acid | Citric acid | HOOCCH$_2$C(OH)(COOH)CH$_2$COOH | Lemons and other fruits | Fruits are citrus |
| 2-hydroxybutane-1,4-dioic acid | Malic acid | HOOCCH$_2$CH(OH)COOH | Many fruits | Malus is Latin name for apple tree |

## Structure of carboxylic acids

Using R as an alkyl group the structure is:

$$R-C\overset{\displaystyle O}{\underset{\displaystyle O-H}{\phantom{|}}}$$

**Link**

Section 1.7

Carboxylic acids are acidic because they release H$^+$ ions when added to water. The pH of these solutions shows that they are weak acids so that the ionisation needed to produce H$^+$ is an equilibrium.

$$RCOOH \rightleftharpoons RCOO^- + H^+$$

### Study point

Water is needed for the acid to produce H$^+$ ions but it is rarely included in the equilibrium equation.

**Worked example**

The pH of a solution of ethanoic acid is 2.52. Calculate the concentration of H$^+$ in the solution.

**Answer**

pH = $-\log_{10}[H^+]$
$[H^+] = 10^{-pH} = 3.02 \times 10^{-3}$ mol dm$^{-3}$

The concentration of the ethanoic acid is actually 0.524 mol dm$^{-3}$. Explain why the concentration of H$^+$ is so much smaller than your original answer.

**Answer**

Ethanoic acid is only a weak acid. When it dissolves in water only a small percentage of H$^+$ ions are formed.

You will learn how to calculate the pH of weak acids accurately in year 13.

# Reactions of carboxylic acids

## 1 As an acid

Carboxylic acids react with bases, carbonates and hydrogencarbonates in a similar way to the reactions of inorganic, strong acids.

### (i) Bases

Both soluble bases, i.e. alkalis, and solid bases, such as metal oxides, neutralise carboxylic acids.

General equation:     acid + base $\longrightarrow$ salt + water

Examples

The alkali sodium hydroxide:

$$CH_3COOH(aq) + NaOH(aq) \longrightarrow CH_3COONa(aq) + H_2O(l)$$
ethanoic acid                                          sodium ethanoate

Since all reagents and products are colourless and soluble in water no **visible** change occurs.

The solid base copper(II) oxide:

$$2HCOOH(aq) + CuO(s) \longrightarrow (HCOO)_2Cu(aq) + H_2O(l)$$
methanoic acid                    copper(II) methanoate

In this reaction, when warmed, the black solid, copper(II) oxide dissolves to form a blue solution of the salt.

### (ii) Carbonates and hydrogencarbonates

Both carbonates and hydrogencarbonates react in a similar way.

General equation:    acid + carbonate $\longrightarrow$ salt + carbon dioxide + water

Examples

$$ZnCO_3(s) + 2CH_3CH_2COOH(aq) \longrightarrow (CH_3CH_2COO)_2Zn(aq) + CO_2(g) + H_2O(l)$$

$$NaHCO_3(aq) + CH_3COOH(aq) \longrightarrow CH_3COONa(aq) + CO_2(g) + H_2O(l)$$

Whether the carbonate or hydrogencarbonate is added to the aqueous acid as a solid or in aqueous solution, carbon dioxide gas is produced. This means that an effervescence is seen and the gas can be shown to be carbon dioxide by testing it with lime water.

## 2 Esterification

Carboxylic acids react with alcohols.

General equation:   carboxylic acid + alcohol $\rightleftharpoons$ ester + water

This reaction is catalysed by the presence of concentrated sulfuric acid. One example is the esterification of ethanoic acid:

$$CH_3COOH + C_2H_5OH \rightleftharpoons CH_3COOC_2H_5 + H_2O$$
ethanoic acid      ethanol          ethyl ethanoate

An easy way to obtain the correct formula for the ester is to draw the OHs of the acid and the alcohol next to each other. Then draw a box round the water to be removed and join the other parts of the molecules together.

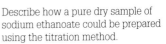
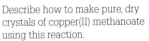

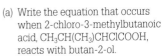

**13** **Knowledge check**

Why can the ester be distilled off leaving the other organic substances in the reaction flask?

**Practical check** 》》

The preparation of esters and their separation is a practical task.

**14** **Knowledge check**

An unknown pure liquid A contains a single alcohol. Using your knowledge of analytical tests in this chapter, outline a simple procedure to allow you to determine whether A is a primary, a secondary or tertiary alcohol.

So the equation for the reaction between methanoic acid and ethanol is:

Methanoic acid            Ethanol                                    Ethyl methanoate

When naming esters, the first part of the name comes from the alcohol and the second part of the name comes from the carboxylic acid. So in the example above the ester is called ethyl methanoate.

This reaction can be used as a test for an alcohol or a carboxylic acid since the esters produced have characteristic sweet, fruity odours.

## Practical activity

If a pure sample of the ester is needed the carboxylic acid, the alcohol and concentrated sulfuric acid are heated together in a flask. This produces an equilibrium mixture of ester with some carboxylic acid and alcohol. In order to have a good yield of ester, the apparatus is set up as shown below and the ester is removed by distillation.

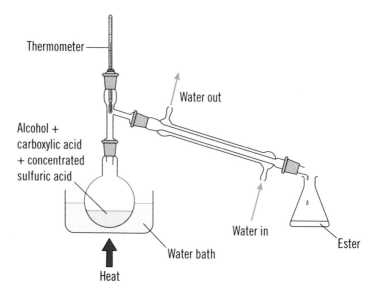

Le Chatelier's principle then says that removal of the ester will pull the equilibrium to the right and more ester is made.

In an exam you need to be able to:

- Draw the apparatus correctly.

- Explain each stage of the method, e.g. why anti-bumping granules are used, why the flask is heated for 15 minutes before the apparatus is reassembled to carry out distillation.

- Apply your knowledge of structure and bonding (Section 1.4) to explain why the ester distils over rather than any remaining alcohol or carboxylic acid.

- Apply your knowledge of equilibria (Section 1.7) to explain why removing the product increases the yield.

# Test yourself

1.  (a) A student was provided with three different alcohols, each of a different classification, but all with the same molecular formula $C_4H_{10}O$.

    Outline a practical procedure, including reagents and conditions, that would allow the student to determine the classification of each alcohol.

    You can assume you are able to isolate any products after each test. Explain your reasoning fully. [6]

    (b) Suggest a name for each type of alcohol. [3]

    (c) What physical property would be different for these isomers. Explain your answer. [3]

2.  (a) Alcohols take part in many reactions.

    Using propan-2-ol as your alcohol, illustrate how it can undergo each of these three reactions: dehydration, oxidation and esterification.

    For each reaction, give reagents and conditions and write an equation showing clearly the structure of the product. [9]

    (b) Draw the apparatus that would be used to obtain the ester as a pure sample. Explain fully why this technique can be used to separate the chemicals present in the reactant mixture. [5]

3.  Ethanol can be produced by the hydration of ethene and by the fermentation of sugars.

    (a) Write an equation for each process. [2]

    (b) State **two** of the essential conditions for fermentation. [2]

    (c) Fermentation is usually the process used to make ethanol for alcoholic drinks but recently it has also been used as a source of biofuel.

    State **two** advantages and **two** disadvantages for the production of biofuel in this way. [4]

# Instrumental analysis

Modern organic chemists will interpret a great deal of molecular data in order to obtain structural information about a compound. A variety of instrumental techniques are used in conjunction with one another, usually to determine the chemical structure of an unknown compound. These techniques can also be automated, which enables molecular data to be interpreted quickly and efficiently. Another benefit is the fact that only a minute amount of a particular compound is required in order to obtain the required data.

## Topic contents

You should be able to demonstrate and apply your knowledge and understanding of:

- The use of mass spectra in identification of chemical structure.
- The use of IR spectra in identification of chemical structure.
- The use of $^{13}C$ and low resolution $^{1}H$ NMR spectra in identification of chemical structure.

### Maths skills >>

- Translate information between graphical and numerical forms.

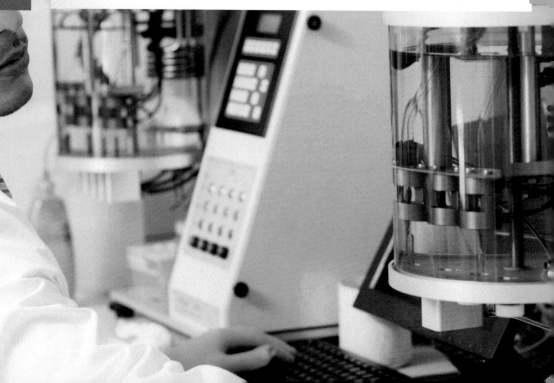

The main methods used in the identification of organic structures are mass spectrometry (MS), infrared spectroscopy (IR), $^{13}$C and $^{1}$H nuclear magnetic resonance spectroscopy (NMR). These techniques will often be used alongside information regarding the elemental composition of the compound, which can be used to calculate empirical and molecular formula.

In this unit, some of the available techniques are considered.

**‹ Link ›**

See Section 1.3 for more information on how a mass spectrometer works.

# Mass spectrometry

The main features of a mass spectrometer are in the diagram:

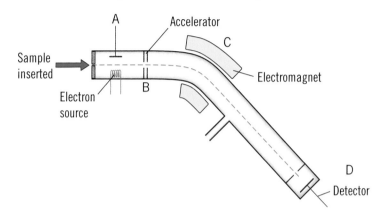

The sample is inserted as a gas and, at A in the mass spectrometer, high-energy electrons knock an electron out of the molecules. This makes positive ions (most of the molecules or atoms lose only one electron to form positive ions with a 1+ charge) that are accelerated by negatively charged plates at B. The positive ions are then deflected by an electromagnet at C and are detected at D. The dotted line shows the path of an ion through the spectrometer. The sample is destroyed during the process. The mass spectrum is usually compared digitally to a spectral database to aid with the identification of the compound.

Bombardment with the high energy electrons means that some molecules are split into smaller parts – fragments. It is important to remember that all of these fragments are also ions. The amount of deflection for each positively charged species depends on the mass – the heavier the ion the less it is deflected. If the strength of the electromagnetic field is altered, species with different masses will pass through the slit and be detected.

This means that a mass spectrum can be produced.

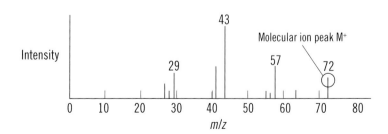

▲ Mass spectrum of pentane

The X-axis shows $m/z$ (mass/charge ratio) and the Y-axis shows the relative abundance, or intensity of each particle. The molecular ion is often shown as M$^+$; it is the peak on the spectrum with the largest $m/z$ and corresponds to the $M_r$ of the compound, as the mass of the lost electron is negligible.

**« Exam tip**

Detailed knowledge of the mass spectrometer is not required; however, an understanding of how to interpret a mass spectrum is vital.

**» Study point**

Imagine rolling a table tennis ball and a ten-pin bowling ball across a table. If someone blows from the side, the lighter table tennis ball will be deflected off its course much more than the heavier ten-pin bowling ball. This is the same as the lighter particles in a mass spectrometer being deflected more by the electromagnetic field.

**Knowledge check**

Complete the following sentences.

In a mass spectrometer ............. ions are produced when an ............. is knocked off. These ions are deflected by a ............. with the ............. being deflected least.

The $M_r$ of the compound is shown in the mass spectrum by ............. .

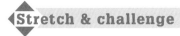
In the mass spectrum on the previous page, the largest $m/z$ is at 72 and therefore the $M_r$ is 72. This value suggests that five carbon atoms are present (mass $5 \times 12 = 60$) with 12 left, so the formula is $C_5H_{12}$. This means that it is an isomer of pentane.

The fragments are then used to give information about which isomer it actually is. $29 = CH_3CH_2^+$, $43 = CH_3CH_2CH_2^+$, $57 = CH_3CH_2CH_2CH_2^+$. This suggests that the compound is pentane $CH_3CH_2CH_2CH_2CH_3$.

In the mass spectrum of compounds containing chlorine or bromine, more than one M⁺ peak will be produced. These, and other fragment peaks, correspond to the presence of more than one isotope for each halogen, i.e. $^{35}Cl$ and $^{37}Cl$ and $^{79}Br$ and $^{81}Br$.

An example is a simplified version of the mass spectrum of 2-chloropropane.

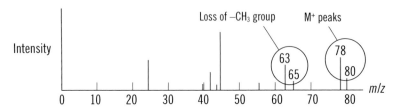

As discussed in Section 1.3, in the spectrum the M⁺ peak at 78 is due to $C_3H_7{}^{35}Cl$ and that at 80 due to $C_3H_7{}^{37}Cl$. There are two molecular ion peaks due to the fact that chlorine has two isotopes. The peaks at 63 and 65 are due to the loss of 15, i.e. $CH_3^+$ from the molecule. The other fragments are caused by rearrangements and are difficult to interpret from the structure of 2-chloropropane.

# Infrared spectroscopy

Radiation in the IR part of the electromagnetic spectrum is absorbed to cause increased vibrations and bending in organic molecules. In an IR spectrometer a range of IR radiation of different energies is passed through the sample and the spectrum produced shows the energies that were absorbed. Since these energies are **characteristic** of the bonds present the **absorptions** can be used to identify which bonds are present. This means that the functional group can be identified and IR can be used in conjunction with mass spectrometry and NMR spectroscopy in order to identify an unknown material.

In any question, use the Data Booklet to match the peaks on the spectrum with the **wavenumbers** of the characteristic absorptions. Some examples are given in the table.

| Bond | Wavenumber/cm⁻¹ |
|---|---|
| C–Br | 500 to 600 |
| C–Cl | 650 to 800 |
| C–O | 1000 to 1300 |
| C=C | 1620 to 1670 |
| C=O | 1650 to 1750 |
| C≡N | 2100 to 2250 |
| C–H | 2800 to 3100 |
| O–H (carboxylic acid) | 2500 to 3200 (very broad) |
| O–H (alcohol / phenol) | 3200 to 3550 (broad) |
| N–H | 3300 to 3500 |

The IR spectrum for ethanol is below.

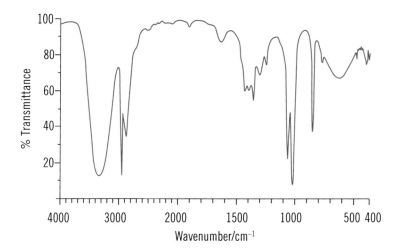

It is not necessary to identify all of the peaks on an IR spectrum, only the ones that can help identify the functional groups present in a compound. The spectrum of ethanol is consistent with the structure of ethanol since it shows an absorption at approximately 1050 cm$^{-1}$ due to C–O and one at approximately 3200 cm$^{-1}$ due to O–H. In the example above the absence of a peak at about 1700 cm$^{-1}$ is also useful since it is always sharp and clear. If seen it shows the presence of C=O.

It would be difficult to positively identify ethanol without some information from another source. The peak assigned to O–H can cause confusion since all organic molecules contain C–H bonds and the absorption due to this is in a very similar range of wavenumbers.

The IR spectra of complex molecules contain many peaks and it would be difficult to interpret these using individual wavenumber values. Databases exist that have the IR spectra for a huge range of molecules and the **whole spectrum** of an unknown compound can be compared with these.

# Nuclear magnetic resonance spectroscopy

Nuclear magnetic resonance (NMR) spectroscopy also corresponds to energy being absorbed to cause a change within molecules. In this case it is to reverse the spin of the **nucleus** of the atom within a **magnetic** field but you are not required to know anything about what actually happens when the energy is absorbed. The absorption of energy causes **resonance** and so it is called nuclear magnetic resonance. The energy absorbed is shown by the **chemical shift**, δ.

Different atoms in a molecule are joined to different atoms/groups and are thus said to be in different **environments.** The environment affects the energy needed to be absorbed to produce the change in the nucleus. This means that the absorption will appear at a different place on the spectrum, and therefore different environments will have different chemical shifts.

The application of NMR technology can be seen in medicine, as it is the basis of MRI (magnetic resonance imaging) analysis and is used as a vital diagnostic tool in hospitals worldwide.

 **Study point**

Whilst data usually refers to absorption peaks they actually appear as troughs, i.e. peaks downwards on the spectra.

**◄◄ Exam tip**

When identifying a compound using IR, link a specific peak and wavenumber to a functional group, e.g. the peak at 1710cm$^{-1}$ is due to C=O.

IR can also be used to analyse and monitor a reaction as, if a new functional group is formed as part of the product, its appearance on the spectrum will be observed, e.g. the formation of a C–Cl bond and therefore a peak at 650–800 cm$^{-1}$ during the free radical substitution by chlorine of a hydrocarbon.

**Knowledge check**

In the IR spectrum of which of the compounds

$$CH_3\overset{\displaystyle O}{\overset{\displaystyle \|}{C}}{-}OCH_3 \quad \text{and} \quad CH_3OCH_3$$

would you expect to see an absorption at 1700 to 1720 cm$^{-1}$?

**▶▶ Key terms**

**Chemical shift** indicates the position for a specific environment on the NMR spectrum – this can be looked up in the data book and linked to a specific part of a molecule.

**Environment** is the nature of the surrounding atoms/groups in a molecule.

**Exam tip** »

Questions will involve interpreting NMR spectra and will not require any knowledge of how they are produced.

# $^{13}C$ spectroscopy

Naturally occurring carbon contains a very small percentage of the isotope $^{13}C$. The presence of atoms of this isotope means that organic compounds will absorb energy and $^{13}C$ NMR spectra can be produced.

This spectrum will have chemical shifts, $\delta$, on the $x$-axis and absorption on the $y$-axis.

An example of a $^{13}C$ spectrum is that for propan-1-ol.

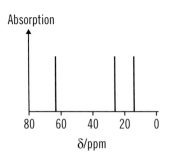
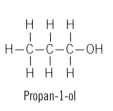

Propan-1-ol

$\delta$ values are needed to interpret the spectrum. This will be provided on the data sheet. Some examples are given in the table.

»» **Study point**

The height of each peak does **not** give any information that is useful in identifying the compound present.

| Type of carbon | Chemical shift, $\delta$ (ppm) | Type of carbon | Chemical shift, $\delta$ (ppm) |
|---|---|---|---|
| $-\overset{\vert}{\underset{\vert}{C}}-\overset{\vert}{\underset{\vert}{C}}-$ | 5 to 40 | $\text{C}=\text{C}$ | 90 to 150 |
| $R-\overset{\vert}{\underset{\vert}{C}}-Cl$ or Br | 10 to 70 | $R-C\equiv N$ | 110 to 125 |
| $R-\overset{\vert}{\underset{\underset{O}{\parallel}}{C}}-\overset{\vert}{C}-$ | 20 to 50 | ⬡ | 110 to 160 |
| $R-\overset{\vert}{\underset{\vert}{C}}-N$ | 25 to 60 | $R-\underset{O}{\overset{\parallel}{C}}-$ (carboxylic acid/ester) | 160 to 185 |
| $-\overset{\vert}{\underset{\vert}{C}}-O-$ | 50 to 90 | $R-\underset{O}{\overset{\parallel}{C}}-$ (aldehyde/ketone) | 190 to 220 |

The spectrum gives two types of information:

- The **number** of different carbon environments – from the number of peaks.
- The **types** of carbon environment – from the chemical shifts.

From the spectrum for propan-1-ol there are three peaks and therefore C atoms in three different environments.

Looking at the chemical shifts to find the environments:

- peak at $\delta$ (C3) = 64 ppm due to **C**–O
- peak at $\delta$ (C2) = 27 ppm due to **C**–C
- peak at $\delta$ (C1) = 15 due to **C**–C.

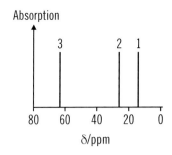

**Knowledge check** 4

Predict the number of peaks and the chemical shifts for each peak on a $^{13}C$ NMR spectrum for:

(a) Ethanol, $C_2H_5OH$

(b) Butanone, $CH_3COCH_2CH_3$.

# $^1$H spectroscopy

This is also called proton ($^1$H) NMR spectroscopy. At AS level you will only be examined on low resolution proton spectroscopy. It is similar to $^{13}C$ spectroscopy in that the spectra give information about:

- The **number** of different proton environments – from the number of peaks.

- The **types** of proton environments – from the chemical shifts.

However, a $^1$H spectrum also gives the ratio of the **numbers of protons** in each environment. This ratio is shown by the relative area/ height of the peaks but is usually quoted in questions.

Typical values for δ are given in the table.

**Study point**

The values for δ on the spectrum can differ considerably so do not be too concerned if your suggested structure would suggest δ values somewhat removed from those that the data sheet gives.

| Type of proton | Chemical shift, δ (ppm) | Type of proton | Chemical shift, δ (ppm) |
|---|---|---|---|
| –CH$_3$ | 0.1 to 2.0 | HC–O | 3.3 to 4.3 |
| R–CH$_3$ | 0.9 | R–OH | 4.5 |
| R–CH$_2$–R | 1.3 | –R=CH–CO | 5.8 to 6.5 |
| CH$_3$–C≡N | 2.0 | ⬡–CH=C | 6.5 to 7.5 |
| CH$_3$–C(=O)\ | 2.0 to 2.5 | ⬡–H | 6.5 to 8 |
| –CH$_2$–C(=O)\ | 2.0 to 3.0 | ⬡–OH | 7.0 |
| ⬡–CH$_3$ | 2.2 to 2.3 | R–C(=O)H | 9.8 |
| HC–Cl or HC–Br | 3.1 to 4.3 | R–C(=O)OH | 11.0 |

**Exam tip**

You must be able to combine information from all of the analytical techniques in this section (MS, IR and NMR) when answering exam questions on deducing organic structures.

**5** **Knowledge check**

For each compound below predict the number of peaks, the ratio of the peak areas and the chemical shifts in a proton NMR spectrum.

(a) $CH_3COCH_3$

(b) $CH_3CH_2CHO$

(c) $(CH_3)_2CHCH_3$

**Study point**

The height of the peaks on $^{13}C$ NMR gives no additional information when trying to deduce the structure of a compound. However, the area/ heights of $^1H$ NMR spectra peaks are important, as they give the ratio of the hydrogen atoms present, relative to one another.

An example of a $^1H$ NMR spectrum is given below.

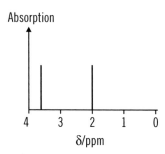

This spectrum is for a compound with molecular formula of $C_3H_6O_2$. The ratio of the area/ height of the peaks is 1:1, therefore the peaks have the same peak area. To identify the compound:

- From the molecular formula there are six protons.

- There are two peaks and therefore two proton environments.

- The ratio of the area/height of peaks is 1:1 so there must be an equal number of protons in each environment, in this case three.

- Using the chemical shifts, the peak at $\delta = 3.6$ matches $O-\textbf{CH}$ and must be $-\textbf{CH}_3$ (numbered 1 in the image below).

  The peak at $\delta = 2.0$ matches $O=C-\textbf{CH}$ and must be $O=C-\textbf{CH}_3$ (numbered 2 in the image below).

The compound is therefore $CH_3COOCH_3$.

$$\begin{array}{ccccccc} & H^2 & O & & H & & \\ & | & \| & & | & & \\ H- & C & -C & -O- & C & -H \\ & | & & & | & & \\ & H & & & H^1 & & \end{array}$$

**Example**

A compound with the molecular formula $C_4H_8O$ has several structural isomers. Use the following spectroscopic data to identify its structure.

Infrared spectrum:

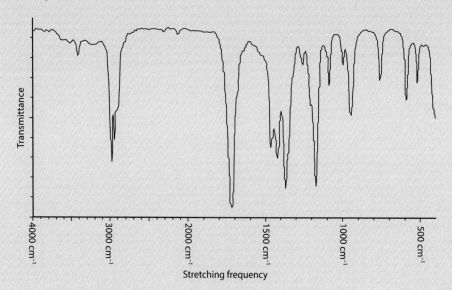

$^{13}$C NMR spectrum:

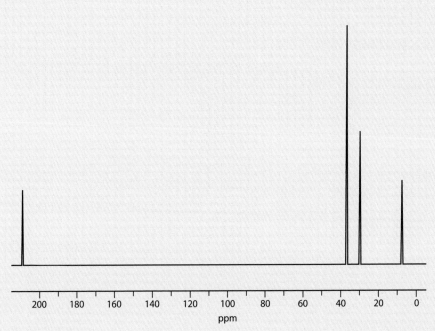

$^{1}$H NMR spectrum:

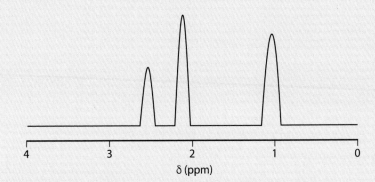

Continued ▶

**Answer**

**Infrared spectrum:**

C=O peak at 1720 cm$^{-1}$

C–H peak at 2800–3100 cm$^{-1}$

This information confirms that the compound must contain a carbonyl group and is therefore either an aldehyde (CHO) or a ketone (COCH$_3$). The compound must therefore be one of the following:

butanal CH$_3$CH$_2$CH$_2$CHO

butanone CH$_3$CH$_2$COCH$_3$

2-methyl propanal CH$_3$(CH$_3$)CHCHO

**$^{13}$C NMR spectrum:**

4 peaks, at 10, 30, 40 and 210 ppm respectively

There are 4 different carbon environments, and therefore 2-methyl propanal can be discounted (as it has only 3 carbon environments).

**$^1$H NMR spectrum:**

3 peaks at 1, 2.2 and 2.6 ppm with peak ratios of 3:3:2 respectively.

Butanal has 4 different hydrogen environments so therefore can be discounted.

The isomer **must be butanone**, as it has 3 hydrogen environments. The $^1$H NMR spectrum confirms this as the chemical shifts and the peak area ratios match the structure.

| Type of proton | Chemical shift (ppm) | Peak ratio |
|---|---|---|
| R–CH$_3$ | 1 | 3 |
| CH$_3$–C(=O) | 2.2 | 3 |
| –CH$_2$–C(=O) | 2.6 | 2 |

# Test yourself

1. What is a molecular ion, and where can it be found on a mass spectrum? [2]

2. What peaks would you expect to be present on a mass spectrum of propanone, $CH_3COCH_3$? [2]

3. List two characteristic IR absorptions for propanone. [1]

4. (a) How would the IR spectrum for propan-2-ol differ from propanone? [1]

   (b) How could these differences in IR absorptions be used to monitor the progress of the oxidation of propan-2-ol into propanone? [2]

5. What functional group is present on the following spectrum? Explain your answer. [2]

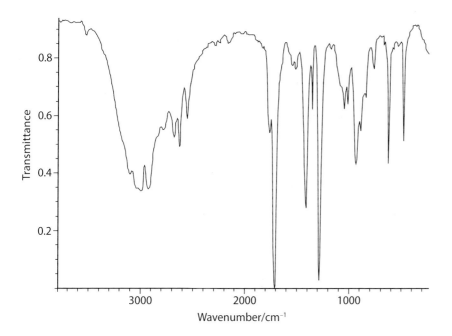

6. How many carbon environments/peaks would be present on a $^{13}C$ NMR spectrum for propanone? [1]

7. State the number of peaks in the $^{13}C$ NMR spectrum for methyl ethanoate. Identify which carbon is responsible for each peak, giving the approximate chemical shift (ppm) of each peak. [3]

8. List the peaks in the $^1H$ NMR spectrum for propanoic acid. Identify which protons are responsible for each peak, giving the approximate chemical shift (ppm) and the relative area of each peak. [3]

9. A compound with an $M_r$ of 58.06 has characteristic IR absorptions at 1700 cm$^{-1}$ and 2800–3100 cm$^{-1}$. It has three $^{13}C$ environments and also three $^1H$ environments with peak area ratios of 3:2:1 with chemical shifts of 0.9, 2.5 and 9.8 ppm respectively. Use this information to deduce the structure of the compound. [1]

# Unit 2

Read the examination questions very carefully. Follow the instructions by giving direct answers to the command words and avoid the temptation to write all you know about a topic.

The example below shows the increase in complexity in a question covering Unit 2 topics. Command words are in **bold**.

Command words are explained on page 8.

The combustion reactions of fossil fuels are exothermic. Two components of petroleum are octane, $C_8H_{18}$, and butane.

The standard enthalpy change of combustion, $\Delta_c H$, of octane is –5512 kJ mol$^{-1}$.

The standard enthalpy change of combustion, $\Delta_c H$, of butane is –2878 kJ mol$^{-1}$.

(a) (i)   **Write** the equation that represents the enthalpy change of combustion of octane.          [1]

    (ii)   **Explain** why the enthalpy change of combustion of octane is negative.          [1]

    (iii)   **Explain** why the enthalpy change of combustion of octane is more negative than that of butane.   [1]

*In part (i), you should recall that only carbon dioxide and water are formed. Since the 'per mole' refers to the octane, to balance the equation you'll need half a mole of oxygen molecules. In part (ii), avoid repeating information given in the question, simply stating 'it is exothermic' does not get credit. Since the command word is 'explain' you should apply your knowledge to explain why it is negative.*

(b) A student carried out a simple method for obtaining the enthalpy change of combustion of octane. He weighed an appropriate amount of octane and used it to heat a specific volume of water. He heated the water for 10 minutes and the temperature of the water rose by 26.0 °C.

**Draw** and label suitable apparatus that the student could use to ensure that his experimental value is as close as possible to the theoretical value.          [3]

*When a diagram of apparatus is required it is clearly not necessary to be perfect, but it should be sufficiently correct as to be of use to a chemist. Since the question states that the 'experimental value is as close as possible to the theoretical value' the apparatus must show that heat loss is kept to a minimum.*

(c) **Explain** the effect on the accuracy of the experimental value obtained if:

    (i)   The water was only heated for two minutes.          [2]

    (ii)   The water was heated for a further two minutes after it had boiled.          [2]

*In this question you are being asked to look at changes to an experiment and draw conclusions. When practical exercises are undertaken, it is important that you think through the implications of the particular actions carried out.*

1 Halogenoalkanes are compounds in which one or more hydrogen atoms in an alkane have been replaced by halogen atoms. Halogenoalkanes have been known for centuries, e.g. chloroethane was produced synthetically in the 15th century. Today they are widely used commercially; however, many have also been shown to be serious pollutants.

(a) Halogenoalkanes can be formed directly from alkanes and alkenes but the ease of formation differs greatly. Briefly outline and explain this difference by considering the types of reactions involved and the bonding in the hydrocarbons. [6 QER]

(No reaction mechanisms are required)

(b) A compound is known to be either 1-chlorobutane or 1-iodobutane. Describe a test to show that the compound is 1-chlorobutane. Give any reagent(s) used and expected observation(s). [3]

(c) 1-Chlorobutane can undergo an elimination reaction with hydroxide ions.
   (i) Draw the displayed formula of the organic product of this reaction. [1]
   (ii) State the conditions required for this reaction. [1]

(d) Explain why 1-chlorobutane has a higher boiling temperature than chloroethane. [2]

(e) CFCs have been shown to be serious pollutants due to their contribution to ozone depletion. Chemists are now replacing them with HFCs. Suggest **two** properties which HFCs should have. [2]

(f) The mass spectrum of a CFC shows two major signals at *m/z* 135 and 137 in the ratio of 3:1 and molecular ion peaks at 170, 172 and 174.

The CFC contains 14.0% carbon, 44.5% fluorine and 41.5% chlorine by mass.

It only has one peak in its $^{13}C$ NMR spectrum.

Use the information to find the structural formula for the CFC. Explain your reasoning. [5]

(Total 20 marks)

*[Eduqas Component 2 2016 Q7]*

2 Pentan-2-ol, $CH_3CH_2CH_2CH(OH)CH_3$, is one of the isomers of $C_5H_{11}OH$. It has a relative molecular mass of 88.1.

A student carried out an experiment on a sample of pentan-2-ol to determine its enthalpy of combustion using the apparatus shown below:

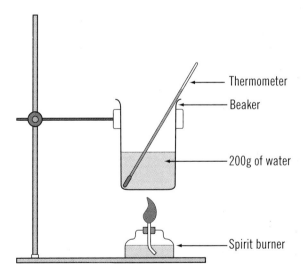

Thermometer
Beaker
200g of water
Spirit burner

The spirit burner was lit and allowed to burn for a few minutes before it was extinguished.

The following results were recorded:

Initial temperature of the water 18.6°C

Final temperature of the water 25.4°C

Initial mass of spirit burner and pentan-2-ol 97.60 g

Final mass of spirit burner and pentan-2-ol 97.42 g

(a) Calculate a value for the enthalpy change of combustion ($\Delta_c H$) of pentan-2-ol in kJ mol$^{-1}$. [4]

(b) (i) Each reading on the thermometer is accurate to ±0.1 °C. Calculate the percentage error in the temperature difference recorded. [1]

(ii) Suggest how, using the same apparatus, the experiment could be improved to reduce this percentage error. Explain your answer. [2]

(c) (i) Suggest an experimental improvement that would reduce error due to heat loss. [1]

(ii) Suggest **one** reason, other than heat loss, why the value obtained for the enthalpy change of combustion is smaller than the theoretical value. [1]

(d) Another isomer of $C_5H_{11}OH$ is 2-methylbutan-2-ol, $(CH_3)_2C(OH)CH_2CH_3$. Standard enthalpy changes of combustion, $\Delta_c H^\ominus$, can be found using standard enthalpy changes of formation, $\Delta_f H^\ominus$.

The table shows some standard enthalpy changes of formation:

| Substance | $\Delta_f H^\ominus$ / kJ mol$^{-1}$ |
|---|---|
| $(CH_3)_2C(OH)CH_2CH_3(l)$ | −380 |
| $O_2(g)$ | 0 |
| $CO_2(g)$ | −394 |
| $H_2O(l)$ | −286 |

(i) Write an equation for the complete combustion of $C_5H_{11}OH$. [1]

(ii) Use the values in the table and the equation in part (i) to calculate the standard enthalpy change of combustion, $\Delta_c H^\ominus$ in kJ mol$^{-1}$, of 2-methylbutan-2-ol. [2]

(iii) State why $O_2(g)$ has a value of zero for its standard enthalpy change of formation, $\Delta_f H^\ominus$. [1]

(e) A student was given separate samples of pentan-2-ol and 2-methylbutan-2-ol but was not told which was which.

Describe a chemical test that the student could use to clearly distinguish between the alcohols. Give any reagent(s) used and expected observation(s) for both compounds. [3]

(f) Another student wanted to make pure 2-chloro-2-methylbutane from 2-methylbutan-2-ol in a multi-step process. One of the steps used was distillation.

(i) An incomplete diagram of the distillation apparatus is shown below:

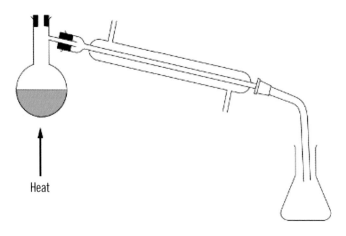

Heat

**Complete the diagram** by drawing a thermometer and clearly labelled arrows to show the flow of water into and out of the condenser. [2]

(ii) The student predicted that since this was a multi-step process the overall yield would be less than 70%.

In the process, 5.00 cm³ of 2-methylbutan-2-ol were used and 4.05 cm³ of 2-chloro-2-methylbutane were made.

Is the student correct? Use the following information to justify your answer:     [3]

| Compound | Density / g cm⁻³ |
|---|---|
| 2-methylbutan-2-ol | 0.805 |
| 2-chloro-2-methylbutane | 0.866 |

$$\text{density} = \frac{\text{mass}}{\text{volume}}$$

(Total 21 marks)

[Eduqas Component 2 2018 Q9]

3  Adam investigated how the initial rate of reaction between hydrochloric acid and magnesium carbonate at 20°C is affected by the concentration of the acid. The equation for the reaction is as follows:

$$MgCO_3(s) + 2HCl(aq) \longrightarrow MgCl_2(aq) + CO_2(g) + H_2O(l)$$

He used 0.50 g of magnesium carbonate and 40 cm³ of 0.20 mol dm⁻³ hydrochloric acid. He measured the volume of carbon dioxide produced at regular time intervals as the reaction proceeded. Part of the apparatus used for the experiment is shown below. The magnesium carbonate was placed in the small glass container which was tipped over to start the reaction and a stopwatch was started at the same time.

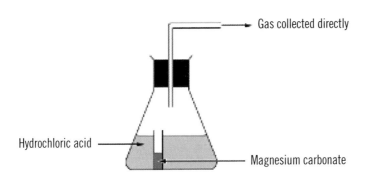

(a) Name the apparatus used to collect and measure the volume of the gas produced.     [1]

(b) Suggest an experimental method other than measuring the volume of gas that would allow the rate of this reaction to be studied.     [2]

(c) Elinor told Adam that in this experiment the carbonate needed to be in excess. Adam replied that it was. Is he correct? Justify your answer.     [2]

(d)  Adam plotted his results as follows:

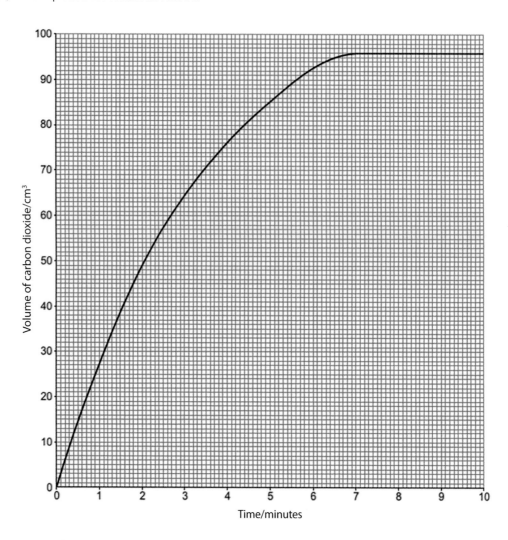

By drawing a tangent to the curve, calculate the initial rate of this reaction. State the unit.          [2]

(e)  Adam then repeated the experiment using 40 cm³ of 0.10 mol dm⁻³ hydrochloric acid. Sketch on the graph in (d) the curve he would expect to obtain. Explain any differences in the curves.          [3]

(f)  State one condition, other than temperature and pressure, which would need to be kept constant in this investigation.          [1]

(Total 11 marks)

*[Eduqas Component 2 2017 Q11]*

# Maths skills

Since the weighting for the assessment of maths skills is a minimum of 20%, it's very important that you've mastered all the maths skills you'll need before sitting your exams. This section gives more explanation and examples of key mathematical concepts you need to understand.

# Arithmetic and numerical computation

## Standard form

Dealing with very large (or small) numbers can be difficult, e.g. the Avogadro constant, an important value in chemistry, is equal to 602 000 000 000 000 000 000 000 $mol^{-1}$. To make such numbers manageable we use standard form.

Standard form is written in the format $A \times 10^n$ so using standard form the Avogadro constant is $6.02 \times 10^{23}$.

To change a number from decimal to standard form:

- Write the non-zero digits with a decimal point after the first number (this is your integer, $A$).
- Count how many places you need to move the decimal point until it is directly to the right of the first number (this is your index number, $n$).
- If the original number was a decimal, your index number is negative.

**Worked example**

Convert 0.0000365 to standard form.

Write the non-zero digits with a decimal point after the first number:

3.65

Next count how many places the decimal point has moved in order to be directly to the right of the first number

0.0000365

The point must be moved 5 times and since the original number is a decimal the index is –5.

Therefore, the standard form is $3.65 \times 10^{-5}$

## Units

To keep numbers more manageable, we use different units for small and large quantities.

Make sure you always give the correct units for your answer – 1 g is very different from 1 kg and make sure you can convert between units.

Often the factor between units is 1000:

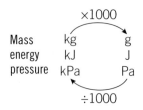

**Worked examples**

**Example 1**

Convert 56 300 J to kJ.

To convert J to kJ you need to divide by 1000:

$$\frac{56\ 300}{1000} = 56.3\ \text{kJ}$$

**Example 2**

Convert 101 kPa to Pa.

To convert kPa to Pa you need to multiply by 1000:

101 × 1000 = 101 000 Pa

The factor between units in volume is also 1000:

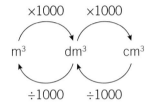

**Worked examples**

**Example 1**

Convert 384 cm³ to m³.

To convert cm³ to dm³ you need to divide by 1000, and to convert dm³ to m³ you need to divide by 1000.

Therefore, to convert cm³ to m³ you need to divide by 1000 × 1000 = 1000 000 or 1 × 10⁶:

$$\frac{384}{1000\ 000} = 0.000384\ \text{or}\ 3.84 \times 10^{-4}\ \text{m}^3$$

**Example 2**

Convert 6.24 m³ to dm³.

To convert m³ to dm³ you need to multiply by 1000:

6.24 × 1000 = 6240 dm³.

> **◀◀ Maths tip**
>
> Dividing by $10^6$ is the same as multiplying by $10^{-6}$.

The difference between the common units of temperature is 273.

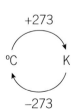

**Worked example**

Convert 35 °C to kelvins.

To convert °C to K you need to add 273:

35 + 273 = 308 K

**‹ Link ›**

Powers are dealt with on page 195

# Calculating units

The equilibrium constant, $K_c$, has variable units. To work out the unit:

- Substitute the units that you know into the equation.
- Cancel out units wherever possible.
- Get rid of any fractions by changing the signs of the powers on the bottom of the fraction.

### Worked examples

#### Example 1

Hydrogen and iodine react together to give hydrogen iodide

$$H_2(g) + I_2(g) \rightleftharpoons 2HI(g)$$

Write the expression for the equilibrium constant, $K_c$, for this reaction and give its unit.

Write the expression

$$K_c = \frac{[HI]^2}{[H_2][I_2]}$$

Substitute the units you know

$$K_c = \frac{(mol\ dm^{-3})^2}{(mol\ dm^{-3})(mol\ dm^{-3})}$$
$$= \frac{(mol\ dm^{-3})(mol\ dm^{-3})}{(mol\ dm^{-3})(mol\ dm^{-3})}$$

Cancel out units where you can. You can cancel $mol\ dm^{-3}$ units from the top and bottom of the fraction.

$$\text{Unit of } K_c = \frac{(\cancel{mol\ dm^{-3}})(\cancel{mol\ dm^{-3}})}{(\cancel{mol\ dm^{-3}})(\cancel{mol\ dm^{-3}})})$$

Therefore, $K_c$ has no unit in this example.

#### Example 2

In the Contact Process, sulfur dioxide reacts with oxygen to give sulfur trioxide:

$$2SO_2(g) + O_2(g) \rightleftharpoons 2SO_3(g)$$

Write the expression for the equilibrium constant, $K_c$, for this reaction and give its unit.

Write the expression

$$K_c = \frac{[SO_3]^2}{[SO_2]^2[O_2]}$$

Substitute the units you know

$$\text{Unit of } K_c = \frac{(mol\ dm^{-3})^2}{(mol\ dm^{-3})^2(mol\ dm^{-3})}$$

Cancel out units where you can. You can cancel $(mol\ dm^{-3})^2$ from the top and bottom of the fraction.

$$\text{Unit of } K_c = \frac{\cancel{(mol\ dm^{-3})^2}}{\cancel{(mol\ dm^{-3})^2}(mol\ dm^{-3})} = \frac{1}{mol\ dm^{-3}}$$

Get rid of the fraction by changing the signs of the powers on the bottom of the fraction

$$\text{Unit of } K_c = mol^{-1}\ dm^3$$

 **Study point**

(Remember 'mol' is the same as 'mol$^1$'.)

# Powers and logarithms

A power tells you how many times you have to multiply a term by itself. Powers can be positive or negative.

Two rules for using powers when dealing with units at AS level are:

$$y^a \times y^b = y^{a+b}$$

This is used when a unit raised to a power is multiplied by the same unit raised to a power, e.g. $(mol\ dm^{-3})^3 = mol\ dm^{-3} \times mol\ dm^{-3} \times mol\ dm^{-3} = mol^3\ dm^{-9}$

$$\frac{1}{y^n} = y^{-n}$$

This is used when a unit is the denominator (bottom number) in a fraction, e.g.

$$\text{Concentration} = \frac{moles}{volume} = \frac{mol}{dm^3} = mol\ dm^{-3}$$

For the Haber Process $N_2(g) + 3H_2(g) \rightleftharpoons 2NH_3(g)$

$$K_c = \frac{[NH_3]^2}{[N_2][H_2]^3}$$

$$\text{Unit of } K_c = \frac{(mol\ dm^{-3})^2}{(mol\ dm^{-3})(mol\ dm^{-3})^3}$$

Cancelling units gives

$$\text{Unit of } K_c = \frac{1}{(mol\ dm^{-3})(mol\ dm^{-3})} = \frac{1}{mol^2\ dm^{-6}} = mol^{-2}\ dm^6$$

At AS level we only use a logarithmic scale in base 10. It is written without a base, e.g. log 1000. It is how many 10s we multiply to get our desired number.

To get 1000 we need $10 \times 10 \times 10$, i.e. $10^3$. Therefore log 1000 = 3.

If the desired number is less than 1 then the log is negative since we find out how many 10s we divide to get our desired number, e.g. for log 0.01

$$\frac{1}{100} = \frac{1}{10^2} = 10^{-2}$$

Therefore log 0.01 = –2

Only one formula uses the log scale: $pH = -\log[H^+]$

This can be rearranged to give $[H^+] = 10^{-pH}$

## Worked examples

An acid has a hydrogen ion concentration of 0.015 mol dm⁻³. Calculate its pH.

$$pH = -\log 0.015 = -(-1.82) = 1.82$$

A solution of nitric acid has a pH of 3.32. Calculate the hydrogen ion concentration of the solution.

$$[H^+] = 10^{-pH} = 10^{-3.32} = 4.79 \times 10^{-4}\ mol\ dm^{-3}$$

 **Maths tip**

Just substitute the [H⁺] value into the formula and use the log button on the calculator.

**Maths tip**

Substitute the pH value into the formula and use the $10^x$ button on the calculator. This is often 'shift log'.

# Handling data

## Using significant figures

Often when you do a calculation, your answer will have many more figures than you need. Using an appropriate number of significant figures will help you to interpret results in a meaningful way. Use the number of significant figures given in the data as a guide for how many you need in the answer.

Remember that:

- The first significant figure is the first figure that is not zero
- Zeros that come after the first significant figure count
- Round up the final significant figure if the next figure is 5 or above

**Maths tip**

Always give your answer to the lowest number of significant figures given in the question. In the example the number of moles is given to 2 sig. figs and the Avogadro constant to 3 sig. figs. Therefore, the answer has to be to 2 sig. figs.

**Worked examples**

65 247 to three sig. figs. is 65 200

0.0031015926 to four sig. figs is 0.003102

Calculate the number of molecules in 3.0 moles of carbon dioxide.

Number of molecules = no of moles × Avogadro constant

$= 3.0 \times 6.02 \times 10^{23} = 18.06 \times 10^{23} = 1.8 \times 10^{24}$

## Arithmetic means

In practical chemistry a mean is an average of your repeated results. To calculate the mean, discount any anomalous results, add up all the concordant results and divide by the number of these results.

e.g. For example:

| Titration | 1 | 2 | 3 | 4 |
|---|---|---|---|---|
| Final reading / $cm^3$ | 24.95 | 24.95 | 25.45 | 24.75 |
| Initial reading / $cm^3$ | 0.00 | 0.35 | 0.80 | 0.20 |
| Titre / $cm^3$ | 24.95 | 24.60 | 24.65 | 24.55 |
| Concordant titres | ✗ | ✓ | ✓ | ✓ |

**Practical check**

For a titration, concordant results are those that are within 0.20 $cm^3$ of each other.

$$\text{Mean titre} = \frac{24.60 + 24.65 + 24.55}{3} = 24.60\ cm^3$$

# Algebra

## Equations

You'll have to learn a number of equations for the AS level exams. All of these have been given in the book, but this page summarises them for you.

### *Amount of substance*

number of particles = number of moles × Avogadro constant

for solids

$$\text{number of moles} = \frac{\text{mass of substance}}{\text{relative formula mass}} \quad \text{or} \quad n = \frac{m}{M_r}$$

for solutions

number of moles = concentration (mol dm$^{-3}$) × volume (dm$^3$) or $n = cv$

for gases

$$\text{number of moles} = \frac{\text{volume (dm}^3)}{\text{molar volume (dm}^3)} \quad \text{or} \quad n = \frac{v}{v_m}$$

$pV = nRT$

Unit of $p$ = Pa, $V$ = m$^3$, $T$ = K

$\dfrac{p_1 V_1}{T_1} = \dfrac{p_2 V_2}{T_2}$    where 1 = initial conditions and 2 = final conditions unit of $T$ = K

For a reaction

$$\% \text{ atom economy} = \frac{\text{total } M_r \text{ of the desired product} \times 100}{\text{total } M_r \text{ of the reactants}}$$

$$\% \text{ yield} = \frac{\text{actual mass (or moles) of product obtained} \times 100}{\text{theoretical mass (or moles) of product}} \times 100$$

For an equilibrium reaction

$aA + bB \rightleftharpoons cC + dD$

$K_c = \dfrac{[C]^c[D]^d}{[A]^a[B]^b}$    [C] = concentration of C in mol dm$^{-3}$

             a = no of moles of a

pH

$pH = -\log[H^+]$

$[H^+] = 10^{-pH}$

### *Energy changes*

To calculate the energy change when a particle absorbs light of a certain wavelength use

$\Delta E = hf$ and $f = \dfrac{c}{\lambda}$

Unit of $E$ = J, $f$ = Hz, $\lambda$ = m

Ionisation energy (kJ mol$^{-1}$) = $\Delta E \times N_A$

**≪ Maths tip**

For no. of moles in solution since the volume will be in cm$^3$ don't forget to change into dm$^3$ by dividing by 1000.

**≫ Study point**

For no. of moles of a gas, only use this expression if the pressure is 1 atm and the temperature is 0 °C or 25 °C.

**≪ Exam tip**

All relevant constants such as $V_m$, $R$, $h$, etc, will be given in the data booklet.

**≪ Exam tip**

Don't forget to convert J to kJ by dividing by 1000.

To calculate the standard enthalpy change of a reaction from experimental data use

$$q = mc\,\Delta T$$

Unit of $q$ = J, $m$ = g

and $\Delta H$ (kJ mol$^{-1}$) $= \dfrac{-q}{n}$

Finally

Enthalpy change of reaction = total energy required to break bonds (reactants) − total energy released in forming bonds (products)

Or $\Delta H = \Sigma$(bonds broken) − $\Sigma$(bonds formed)

# Changing the subject of an equation

You'll often need to rearrange an equation to make the quantity that you've been asked to calculate the subject, i.e. the term on its own on one side of the equation.

Always remember whatever you do to one side of the equation you must do to the other side.

You'll often need to use more than one equation to answer a question.

**Worked examples**

**Example 1**

For the reaction

$$H_2(g) + I_2(g) \rightleftharpoons 2HI(g)$$

the equilibrium constant, $K_c$, is 60.0 at 450 °C.

If the equilibrium concentration of hydrogen iodide is 20.0 mol dm$^{-3}$ and that of hydrogen is 2.50 mol dm$^{-3}$, calculate the equilibrium concentration of iodine at that temperature.

$$K_c = \frac{[HI]^2}{[H_2][I_2]}$$

You need to rearrange it to make $[I_2]$ the subject. To get rid of $[I_2]$ from the bottom, you need to multiply by $[I_2]$. Don't forget you need to multiply both sides by $[I_2]$.

$$[I_2]K_c = \frac{[HI]^2[I_2]}{[H_2][I_2]}$$

You can cancel out $[I_2]$ on the right-hand side.

$$[I_2]K_c = \frac{[HI]^2}{[H_2]}$$

To get $[I_2]$ on its own divide by $K_c$. Again, remember you must divide both sides by $K_c$.

$$\frac{[I_2]K_c}{K_c} = \frac{[HI]^2}{[H_2]K_c}$$

You can cancel out $K_c$ on the left-hand side.

$$[I_2] = \frac{[HI]^2}{[H_2]K_c}$$

You can now substitute the values given in the question into the equation.

$$[I_2] = \frac{20.0^2}{(2.50)(60.0)} = 2.67 \text{ mol dm}^{-3}$$

### Example 2

Calculate the mass of sodium hydroxide that is needed to make 200 cm$^3$ of 0.40 mol dm$^{-3}$ sodium hydroxide.

For solids, $n = \dfrac{m}{M_r}$

To get m on its own, multiply both sides by $M_r$ and cancel.

$$n \times M_r = \frac{m}{\cancel{M_r}} \times \cancel{M_r}$$

To calculate the number of moles of sodium hydroxide use $n = c \times v$

$$n = 0.40 \times \frac{200}{1000} = 0.080$$

Therefore, mass = $0.08 \times 40.01 = 3.2$ g

**Maths tip**

Since all the numbers in the question are given to 3 sig.figs., give your answer to 3 sig.figs.

**Maths tip**

Remember you need to divide the volume by 1000 to change cm$^3$ to dm$^3$.

# Graphs

Graphs are one of the most important tools available to scientists to display data.

Normally at least half the marks on a graph question are awarded for presenting your data. Therefore, it is important that you:

- Choose a suitable scale – use over half the graph paper.

- Label your axes and give units.

- Ensure you have your graph the right way round. Your x axis (horizontal axis) should show the variable you are changing (independent variable). Your y axis (vertical axis) should show the variable you are measuring (dependent variable).

- Do not join your points dot-to-dot. Normally in a graph we are seeking the trend that the results show.

- Draw a line of best fit. This should pass through or near as many points as possible – you can ignore any anomalous points. The best-fit line can be straight or a smooth curve.

Graphs can be used to measure the rate of a reaction by finding the gradient of the best-fit line. If it is a straight line, pick two points on the line that are easy to read. Draw a vertical line from one point (change in y) and a horizontal line from the other point (change in x).

$$\text{Gradient} = \frac{\text{change in } y}{\text{change in } x} = \frac{\Delta y}{\Delta x}$$

If the best-fit line is a curve, then to find the gradient at a certain point, you have to draw a tangent to the curve at that point.

To do this place a ruler at the required point on the curve and position the ruler so that you can see all of the curve. Adjust the ruler until the space between the ruler and the curve is equal on both sides of the point. Draw a line along the ruler to make the tangent and extend the line across the graph.

**Worked examples**

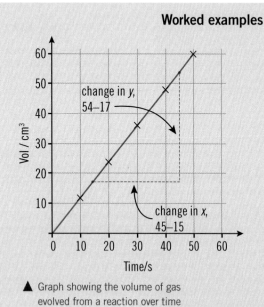

▲ Graph showing the volume of gas
evolved from a reaction over time

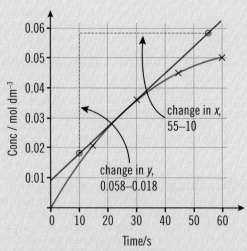

▲ Graph showing the concentration of a
product against time

### Example 1: Straight line graph

$$\text{Gradient} = \frac{\text{change in } y}{\text{change in } x} = \frac{37}{30} = 1.2 \text{ cm}^3\text{s}^{-1}$$

### Example 2: Curved line graph

The rate of the reaction at 25 seconds is equal to the gradient of the tangent to the curve at 25 seconds:

$$\text{Gradient} = \frac{\text{change in } y}{\text{change in } x} = \frac{0.040}{45} = 8.9 \times 10^{-4} \text{ cm}^3\text{s}^{-1}$$

**Maths tip** ≫

A tangent is a straight line that touches but does not cut into a curve.

# Geometry

You'll have to be able to draw 2D representations of 3D shapes by using wedges and dotted lines.

For example, the shape of a methane molecule:

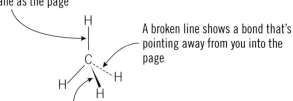

An ordinary line shows a bond that's in the same plane as the page

A broken line shows a bond that's pointing away from you into the page

A wedge shows a bond that's pointing towards you out of the page

# Knowledge check answers

## Unit 1

### 1.1

1 (a) P = 4, O = 10
 (b) Al = 2, O = 6, H = 6

2 27

3 (a) Sodium sulfate
 (b) Calcium hydrogencarbonate
 (c) Copper(II) chloride

4 (a) $Al_2O_3$
 (b) $K_2CO_3$
 (c) $(NH_4)_2SO_4$

5 (a) −3
 (b) 0
 (c) +7
 (d) +6

6 (a) $2SO_2 + O_2 \longrightarrow 2SO_3$
 (b) $Fe_2O_3 + 3CO \longrightarrow 2Fe + 3CO_2$
 (c) $2Al + 6HCl \longrightarrow 2AlCl_3 + 3H_2$
 (d) $8HI + H_2SO_4 \longrightarrow 4I_2 + 4H_2O + H_2S$

7 $Ba^{2+}(aq) + SO_4^{2-}(aq) \longrightarrow BaSO_4(s)$

### 1.2

1 Cu-63: 29 p, 29 e, 34 n
 Cu-65: 29 p, 29 e, 36 n

2 (a) 53 p, 54 e
 (b) 12 p, 10 e

3 $^{234}Pa$

4 $^{83}_{38}Sr \longrightarrow {}^{83}_{37}Rb + \beta^+$

5 32 days

6 15 minutes

7 (a) (i)

| ↑↓ | ↑↓ | ↑↓ | ↑ | ↑ | ↑↓ | ↑ | ↑ | ↑ |
|---|---|---|---|---|---|---|---|---|
| 1s | 2s | | 2p | | 3s | | 3p | |

 (ii)

| ↑↓ | ↑↓ | ↑↓ | ↑↓ | ↑↑ |
|---|---|---|---|---|
| 1s | 2s | | 2p | |

 (b) $1s^2 2s^2 2p^6 3s^2 3p^6 3d^5 4s^1$

8 9 (1 s, 3 p and 5 d)

9 First ionisation energy of phosphorus higher since the electron repulsion between the two paired electrons in one p orbital in sulfur makes one of the electrons easier to remove. There are no paired electrons in phosphorus.

10 The element belongs to Group **2** in the periodic table because there is a **large jump** between the **second** and **third** ionisation energies.

11 (a) Line with $6.9 \times 10^{14}$ Hz has the higher energy since $E \propto f$
 (b) Line with $4.6 \times 10^{14}$ Hz has the higher wavelength since $f \propto 1/\lambda$

12 Absorption spectrum produced when atoms absorb energy while emission spectrum produced when atoms release energy.
 Absorption spectrum comprises dark lines while emission spectrum comprises coloured lines.

13 C

14 $495\,kJ\,mol^{-1}$

### 1.3

1 (a) 74.12
 (b) 248.3

2 Vaporised sample is bombarded with a stream of electrons (knocking an electron out of the sample).

3 1.008

4 The molecular ion, $H_2^+$, is not stable and splits to give a hydrogen atom and an $H^+$ ion.

5 (a) 23.0 g
 (b) 1533 g

6 (a) 2.12 g
 (b) 0.0136 mol

7 1.72 g

8 $FeCl_3$

9 $0.816\,dm^3$

10 Pressure in Pa
 Volume in $m^3$
 Temperature in K

11 $3.15\,dm^3$

12 $338\,cm^3$

13 $2.50 \times 10^{-3}\,mol\,dm^{-3}$

14 $0.148\,mol\,dm^{-3}$

15 $0.0450\,mol\,dm^{-3}$

16 45.8%

17 0.93 %

18 (a) 4
 (b) 4
 (c) 1
 (d) 3
 (e) 5

### 1.4

1 ionic, covalent, coordinate

2 MgO

3 (a) Linear, trigonal planar, tetrahedral, trigonal bipyramidal.
 (b) 180°, 120°, 109.5°, 90° and 120°, 90°.
 (c) The shapes are different. $BCl_3$ has three pairs of electrons, all of which are bonding so its shape is trigonal planar. $PCl_3$ has four pairs of electrons, three bonding pairs and one lone pair, so its shape is trigonal pyramidal. Since the molecules contain a different number of electron pairs, which arrange themselves to be as far apart as possible, so that the repulsion between them is at a minimum, the shapes are different.

**1.5**

1. Caesium chloride contains caesium ions, $Cs^+$, and chloride ions, $Cl^-$. Molecules are formed when two or more atoms bond together.

2. (a) Graphite contains delocalised electrons which can move along the layers so an electric current can flow. In diamond all the outer electrons are held in localised bonds.

   (b) The weak forces between the layers are easily broken, so the layers can slide over each other.

3. Simple molecular structure since it cannot conduct electricity at all. (Simple molecular structures can be soluble in water due to hydrogen bonding.)

4. (a) (i) Metallic  (ii) Ionic.

   (b) The layers of cations in the lattice slide over each other. Delocalised electrons between the cations prevent forces of repulsion forming between the layers. Thus, aluminium keeps its shape.

**1.6**

1. 

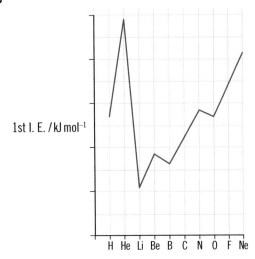

2. Fluorine and nitrogen are in the same period, but fluorine has more protons in its nucleus so there is a greater attraction between the nucleus and the bonding electrons.

3. The Ca has lost 2 electrons and has been oxidised. The H has gained an electron and has been reduced.

4. Sr has been oxidised; oxidation number has increased from 0 to +2. H has been reduced; oxidation number reduced from +1 to 0.

5. Calcium carbonate.

6. Add aqueous sodium hydroxide to both solutions. The magnesium chloride solution will form a white precipitate but there will be no observable change with the barium chloride solution. (Or add aqueous sodium sulfate to both solutions. The barium chloride solution will form a white precipitate but there will be no observable change with the magnesium chloride solution.)

7. There is a change in colour from orange to brown. It is a redox reaction since bromine changes from oxidation number 0 to –1 so is reduced and iodide changes from oxidation number –1 to 0 so is oxidised.

8. (a) Benefit: Kills bacteria / microbes. Risk: Toxic in high concentration

   (b) Benefit: Reduces tooth decay. Risk: Can cause fluorosis.

**1.7**

1. (a) No effect. There are 2 moles of gas on each side of the equation.

   (b) Moves to the right. Moves in the endothermic direction. The reaction is endothermic from left to right.

2. (a) $K_c = \dfrac{[SO_3]^2}{[SO_2]^2\,[O_2]}\ dm^3\,mol^{-1}$

   (b) $K_c = \dfrac{[PCl_3]\,[Cl_2]}{[PCl_5]}\ mol\,dm^{-3}$

3. 50

4. Nitric acid fully dissociates in aqueous solution. The aqueous hydrogen ion concentration is equal in magnitude to the concentration of the acid.

5. A strong acid is one that fully dissociates in aqueous solution while a concentrated acid is one that consists of a large quantity of acid and a small quantity of water.

6. (a) 2

   (b) $3.16 \times 10^{-3}\,mol\,dm^{-3}$

7. Potassium nitrate and water

8. It is not stable (in air) since it reacts with atmospheric carbon dioxide and it has a low molar mass.

9. (a) A measuring cylinder is not accurate enough so the exact amount of moles of base would not be known / the percentage error from the equipment would be too large

   (b) To show the end point / when to stop adding acid

# Unit 2

**2.1**

1. This is the enthalpy change when one mole of a substance is formed from its constituent elements in their standard states under standard conditions.

2. Oxygen gas is an element in its standard state.

3. $-235\,kJ\,mol^{-1}$

4. $-313\,kJ\,mol^{-1}$

5. $-484\,kJ\,mol^{-1}$

6. $243\ kJ\,mol^{-1}$

7. $83.6\,kJ$

8. $-54.3\,kJ\,mol^{-1}$

9. (a) Rate of reaction is quicker.

   (b) Extrapolation gives the temperature that would have been reached if the reaction occurred instantly / extrapolation allows for heat loss during the experiment.

**2.2**

1. $0.85\ cm^3\,s^{-1}$

2. $0.004\,mol\,dm^{-3}\,s^{-1}$

3. There is an increase in concentration of acid therefore there are more molecules in a given volume, so there is an increase in the number of collisions per unit time. This means that there is a greater chance that the number of effective collisions will increase.

4. $-52\,kJ\,mol^{-1}$

5. Only the molecules with an energy equal to or greater than the activation energy are able to react. At the higher temperature, the mean kinetic energy of the acid molecules increases and many more molecules have sufficient energy to react.

6 A catalyst is a substance that increases the rate of a chemical reaction without being used up in the process. It increases the rate of reaction by providing an alternative route of lower activation energy.

7 Measure the volume of carbon dioxide produced (using a gas syringe) at constant time intervals / measure the change in mass at various times (using weighing scales).

8 Some of the hydrogen gas will escape before the bung is properly replaced.

9 (a) The total volume affects the concentrations of reactants and must remain the same to make a fair comparison between runs.

   (b) Rates vary rapidly with changes in temperature.

   (c) Once the peroxide is added the reaction starts (accept the peroxide is the oxidising agent).

## 2.3

1 Hydrogen

## 2.4

1 The longest carbon chain has 8 carbon atoms. Name based on 'oct'.

2 (a)

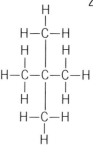

   (b)

3 (a) 3-methylpent-1-ene.

   (b) 2-bromo-2-chloropropan-1-ol.

4 (a) $C_4H_7ClO$

   (b)

   (c)

5 Alkenes have similar **chemical** properties but show a trend in **physical** properties because they all belong to the same **homologous series**.

6 $C_{100}H_{202}$.

7 Ratio number of moles $C:H:O = \dfrac{40.00}{12.0} : \dfrac{6.67}{1.01} : \dfrac{53.33}{16.0}$

   $= 3.33 : 6.67 : 3.33$

   Divide by smallest        $= 1:2:1$

   Empirical formula         $= CH_2O$

   Relative empirical mass $= 12 + 2 + 16 = 30$

   $M_r$ = approx. 55    Molecular formula $= C_2H_4O_2$

8 0.660g of $CO_2$ contain $0.660 \times \dfrac{12}{44}$ g of carbon $= 0.18$g.

   0.225g of $H_2O$ contain $0.225 \times \dfrac{2}{18}$ g of hydrogen $= 0.025$g

   Ratio $C:H = \dfrac{0.180}{12.0} : \dfrac{0.025}{1.01}$

   $= 0.015 : 0.0248$

   Divide by smaller $= 1:1.65$

   $= 3:5$

   Empirical formula $= C_3H_5$

   Relative empirical mass $= 41$.

   $M_r$ given as approximately 80. Molecular formula $= C_6H_{10}$.

9 (a)

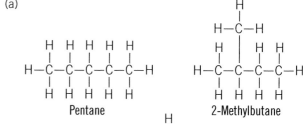

Pentane          2-Methylbutane

2,2-Dimethylpropane

   (b)

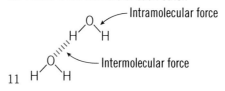

Pent-1-ene      Pent-2-ene

10 A and B are isomers of each other.

Intramolecular force

Intermolecular force

11

## 2.5

1 Two of water, carbon dioxide and methane.

2 The presence of **nitric** acid and **sulfuric** acid in rain water **lowers** the pH.

3 $CaCO_3(s) + 2HNO_3(aq) \longrightarrow Ca(NO_3)_2(aq) + CO_2(g) + H_2O(l)$

4 Both chemicals consist of molecules which have van der Waals forces between them. The molecules in octane are bigger and have more electrons and so the van der Waals forces between the molecules are stronger. This means it takes more energy to separate the octane molecules than to separate the methane molecules.

5 The smoke is carbon – this forms from incomplete combustion of fuel.

6 $C_4H_{10} + 4\frac{1}{2}O_2 \longrightarrow 4CO + 5H_2O$

7 $C_8H_{18} + 12\frac{1}{2}O_2 \longrightarrow 8CO_2 + 9H_2O$

8 Alkanes are generally unreactive because **they are non polar and do not contain multiple bonds.** Radicals are very reactive because they **have an unpaired electron.**

9 $CH_4 + 4Cl_2 \longrightarrow CCl_4 + 4HCl$

10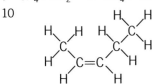

11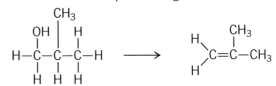

12 Z isomer. Cl has a greater $A_r$ than C.

13 (a)

Br—C—Br with Br Br on top and H H on bottom

H—C—C—H with Br Br on top and Br Br on bottom

(b) They are the same compound.

14 $(CH_3)_2C{=}CH_2 + HBr \longrightarrow (CH_3)_2CBrCH_3$.

15 1 Bromine water is brown and therefore easier to show loss of colour.

2 Bromine water is a liquid and easier to handle than the gas chlorine.

16 Poly(1-bromo-2-nitroethene).

17

$$-C-C-C-C-C-C-$$ with Br NO$_2$ Br NO$_2$ Br NO$_2$ on top and H H H H H H on bottom

18 $CH_2$.

19

Br—C=C—CO$_2$CH$_3$ (with H, H substituents)

**2.6**

1 3-bromo-1,1-dichloro-1-fluorobutane.

2 2-methyl-2-halogenobutane and sodium hydroxide.

3 Aqueous conditions are required and heat.

4

$$H-C-C-C-C-H \quad + \quad NaOH$$ (with CH$_3$ branch, H, Cl, H, H substituents)

↓

$$H-C-C-C-C-H \quad + \quad NaCl$$ (with CH$_3$ branch, H, OH, H, H substituents)

5 Chlorine is smaller than bromine and there is less shielding, so the nucleus exerts more attraction on the bonding pair of electrons.

6 Chlorine is smaller than bromine. The chlorine atom can get closer to the carbon atom and therefore the attraction between the shared pair of electrons and the nuclei is greater.

7 Warm the suspected bromocompound with aqueous sodium hydroxide. Neutralise the excess sodium hydroxide with dilute nitric acid. Add aqueous silver nitrate. If the compound is a bromoalkane a cream precipitate that is soluble in concentrated ammonia is formed.

8 $Ag^+(aq) + I^-(aq) \longrightarrow AgI(s)$

9 (a) 2-methylbutan-1-ol.

(b) 2-methylbut-1-ene.

10 (a) 2-methylbutan-2-ol.

(b) 2-methylbut-1-ene and 2-methylbut-2-ene.

11 Fluorine is smaller than chlorine. The fluorine atom can get closer to the carbon atom and therefore the attraction between the shared pair of electrons and the nuclei is greater.

**2.7**

1 Ethanol could be removed by cooling the gaseous mixture. Ethanol and steam will liquefy. The liquid mixture can be separated by fractional distillation.

2 Fractional distillation is needed as the boiling temperatures of water and ethanol are quite close together.

3

$$H-C-C-C-H \longrightarrow \text{(alkene)}$$ (left: OH, H, CH$_3$, H substituents; right: H$_2$C=C(CH$_3$) with CH$_3$)

4 (a)

(E) but-2-ene (CH$_3$ and H / H and CH$_3$)

(Z) but-2-ene (CH$_3$ and CH$_3$ / H and H)

$$H-C{=}C-C-C-H$$ (But-1-ene, with H H H H and H H substituents)

But-1-ene

(b) No elimination products are possible.

5 (a) Primary.

(b) Tertiary.

(c) Secondary.

(d) Primary.

6 $CH_3CH_2CH_2OH + [O] \longrightarrow CH_3CH_2CHO + H_2O$

7 $CH_3CH_2CH_2CH_2OH + 2[O] \longrightarrow CH_3CH_2CH_2COOH + H_2O$

8 Apparatus suitable for reflux. You need to be able to heat the liquids for a time without losing them through evaporation.

9 • Ethanoic acid in burette (accept sodium hydroxide)
• Measure volume of sodium hydroxide into conical flask using a pipette
• Use an indicator
• Titrate until colour of indicator just changes
• Read burette
• Repeat titration without indicator – using exactly the same volumes
• Evaporate part of the solution, leave to crystallise
• Filter and dry crystals

10 • Measure methanoic acid into a beaker, e.g. 25cm$^3$
   • Heat acid gently
   • Add excess copper(II) oxide – stirring to ensure all the acid reacts
   • Remove unreacted copper(II) oxide by filtration
   • Evaporate part of the solution, leave to crystallise
   • Filter and dry crystals

11 (a) $CaCO_3(s) + 2CH_3COOH(aq)$
   $\longrightarrow (CH_3COO)_2Ca(aq) + CO_2(g) + H_2O(l)$
   (b) $KHCO_3(aq) + HCOOH(aq)$
   $\longrightarrow HCOOK(aq) + CO_2(g) + H_2O(l)$

12 (a)

   (b) Sulfuric acid acts as a catalyst.

13 The ester has the lowest boiling temperature. Alcohols, carboxylic acids and water all contain the –OH group and can hydrogen bond. This increases the boiling temperatures. Esters cannot hydrogen bond.

14 Place 1 cm$^3$ of A in a test tube and add 3cm$^3$ of acidified potassium dichromate, and heat in a water bath for 5 minutes. (Any suitable amounts and timescale accepted.)

   If the solution does not change from orange to green, then A is a tertiary alcohol and so no more testing is required.

   If the colour does change, separate the product of the oxidation of A – any sensible suggestion, e.g. distillation.

   Test the isolated oxidation product with sodium hydrogencarbonate solution. If fizzing occurs then A was a primary alcohol. If not, A was a secondary alcohol.

## 2.8

1 In a mass spectrometer **positive** ions are produced when an **electron** is knocked off. These ions are deflected by a **magnetic field** with the **heaviest** being deflected least. The $M_r$ of the compound is shown, in the mass spectrum, by **the peak with the largest $m/z$.**

2 (a) 60. This is the molecular ion and therefore the $m/z$ is the $M_r$.
   (b) A fragment with $m/z$ of 17 has been lost. This suggests an OH group has been removed producing a peak at 43 for the $CH_3CHCH_3$ ion.

3

Since this contains –C=O whilst $CH_3OCH_3$ does not
It is C=O that gives an absorption at 1700 to 1720cm$^{-1}$.

4 (a) 2 peaks with δ in range 50 to 90 and in range 5 to 40.
   (b) 4 peaks 1 with δ in range 190 to 210 and 3 with δ in range 5 to 40.

5 (a) 1 peak with δ in range 1.9 to 2.9 for CHC=O
   (b) 3 peaks with peak areas in ratio 3:2:1. $CH_3CH_2CHO$ δ in range 0.7 to 1.6
   $CH_3\textbf{CH}_2CHO$ δ in range 2.0–3.0 $CH_3CH_2\textbf{CHO}$ δ at 9.8
   (c) 2 peaks with peak areas in ratio 9:1. δ for both in range 1.0 to 2.0

# Test yourself answers

## Unit 1

### 1.1

1. $Ra(OH)_2$ [1]
2. $+6$ [1]
3. $Fe_3O_4 + \textbf{4}CO \longrightarrow \textbf{3}Fe + \textbf{4}CO_2$ [1]
4. $C_2H_5OH + \textbf{3}O_2 \longrightarrow \textbf{2}CO_2 + \textbf{3}H_2O$ [1]
5. $Ca + 2H_2O \longrightarrow Ca(OH)_2 + H_2$ [1]
6. $SiCl_4 + 2H_2O \longrightarrow SiO_2 + 4HCl$ [1]
7. $2(NH_4)_3PO_4 + 3Pb(NO_3)_2 \longrightarrow Pb_3(PO_4)_2 + 6NH_4NO_3$ [1]
8. (a) $3NO_2(g) + H_2O(l) \longrightarrow 2HNO_3(g) + NO(g)$ [1]
   (b) $+4$ in $NO_2$, $+5$ in $HNO_3$, $+2$ in $NO$ [2]
9. $Ca^{2+}(aq) + CO_3^{2-}(aq) \longrightarrow CaCO3(s)$ [1]

### 1.2

1. (a)

| Particle | Relative mass | Relative charge |
|----------|---------------|-----------------|
| Proton | 1 | +1 |
| Neutron | 1 | 0 |
| Electron | Negligible | −1 |

   (b) 47 protons, 46 electrons
   (c) (i) Number of protons in the nucleus of an atom [1]
   e.g. Na has 11 protons [1]
   (ii) Atoms that have the same number of protons but different numbers of neutrons / same atomic number but different mass number [1]
   e.g. Cu-63 has 29 protons and 34 neutrons while Cu-65 has 29 protons and 36 neutrons [1]
   (d) Fluorine
   (e) $1s^2\ 2s^2\ 2p^6\ 3s^2\ 3p^6\ 3d^2\ 4s^2$

2. (a) (i) $^{199}_{78}Pt \longrightarrow\ ^{199}_{79}Au +\ ^{0}_{-1}\beta$ (accept $^{0}_{-1}e$)
   (ii) A neutron changes into a proton (and an electron is emitted by the nucleus)
   (iii) 31 minutes
   (b) $^{23}Na$
   (c) Radioactivity damages or destroys the DNA of a cell / causes mutations in a cell [1]
   Most damage is caused by gamma rays if the source is outside the body, but if the source is ingested then alpha particles cause most damage [1]
   Gamma rays are the most penetrating of the radioactive emissions (so can enter the body easier) [1]
   Alpha particles are the most ionising (and cannot leave the body) [1]

3. (a) The element cannot be in Group 1, 2 or 3 [1]
   There is not a large jump before the fourth ionisation energy [1]
   (b) (i) Higher since it has greater nuclear charge [1]
   but little/no extra shielding [1]
   (ii) Higher since nitrogen only has unpaired 2p electrons, oxygen has two unpaired and two paired 2p electrons [1]
   Repulsion between paired electrons makes it easier to remove one of the electrons / takes more energy to remove unpaired electron [1]
   (iii) Higher since nitrogen's outer electron is closer to the nucleus (so stronger attraction to the nucleus) [1]
   and there is less shielding from the inner electrons [1]
   (c) (i) $N^{2+}(g) \longrightarrow N^{3+}(g) + e^-$
   (ii) (All successive ionisation energies increase but) a value greater than the one expected from the trend in increase in successive ionisation energies shows a new subshell [1]
   A large increase in successive ionisation energies shows that an electron has been removed from a new shell closer to the nucleus [1]

4. Atomic spectrum of hydrogen consists of separate series of lines [1]
   Lines get closer together as their frequency increases [1]
   (Can get these points from labelled diagram)
   Lines arise from atom being excited by absorbing energy / electron jumping up to a higher energy level [1]
   Falling back down and emitting energy (in the form of electromagnetic radiation) [1]
   Lines are discrete because energy emitted is equal to the difference between two energy levels [1]

5. (a) $f = 4.58 \times 10^{-19}$ (J)/$6.63 \times 10^{-34}$ (Js) [1]
   $f = 6.91 \times 10^{14}$ ($s^{-1}$) [1]
   $\lambda = 3.00 \times 10^8/6.91 \times 10^{14} = 4.34 \times 10^{-7}$ (m) [1]
   $\lambda = 434$ nm [1]
   (b) Energy would be less since wavelength inversely proportional to energy / wavelength inversely proportional to frequency and frequency directly proportional to energy

### 1.3

1. 24.6 g
2. (a) (i) It must be vaporised [1] and ionised [1]
   (ii) $A_r = \dfrac{(92.21 \times 28) + (4.70 \times 29) + (3.09 \times 30)}{100}$ [1]
   $A_r = 28.11$ [1]
   (iii) This is the average mass of one atom of the element relative to one-twelfth the mass of one atom of carbon-12

(b) Mass number 1 is $^1H^+$, Mass number 18 is $(^1H_2O)^+$, Mass number 20 is $(^2H_2O)^+$  [2]

(1 mark if two out of three correct)

3 (a)

| P | O | Cl |
|---|---|---|
| $\dfrac{20.0}{31.0}$ | $\dfrac{10.4}{16.0}$ | $\dfrac{69.6}{35.5}$ |
| 0.645 | 0.650 | 1.96 |
| 1 | 1 | 3 |

[1]

$POCl_3$  [1]

(b) $M_r$ / number of atoms of any element in the compound

4 (a) Moles $Na_2CO_3 = 0.045 \times 0.025 = 1.125 \times 10^{-3}$  [1]

Moles $HNO_3 = 2 \times 1.125 \times 10^{-3} = 2.25 \times 10^{-3}$  [1]

Concentration $HNO_3 = \dfrac{2.25 \times 10^{-3}}{0.0236} = 0.0953$ mol dm$^{-3}$  [1]

(b) $n = \dfrac{pV}{RT}$  [1]

$n = \dfrac{101\,000 \times 700 \times 10^{-6}}{8.31 \times 301}$  [1]

$n = 0.0283$  [1]

5 (a) Moles $HCl = 1.20 \times 0.020 = 0.024$  [1]

Moles $CaCO_3 = 0.024/2 = 0.012$  [1]

Mass $= 0.012 \times 100 = 1.20$ g  [1]

(b) Moles $CO_2 = 0.012$  [1]

Volume $= 0.012 \times 24.5 = 0.294$ dm$^3$  [1]

(c) $\dfrac{V}{323} = \dfrac{0.294}{298}$  [1]

$V = 0.319$ dm$^3$  [1]

6 (a) Atom economy $= \dfrac{47.9}{189.9 + 48.6} \times 100$  [1]

$= 20.1\%$  [1]

(b) Moles $TiO_2 = \dfrac{447000}{79.9} = 5594.5$  [1]

Mass Ti $= 5594.5 \times 47.9 = 267976$ g $= 268$ kg  [1]

7 Moles $Na_2SO_4 = \dfrac{2.39}{142.1} = 0.0168$  [1]

Moles $H_2O = \dfrac{3.03}{18.02} = 0.168$  [1]

$x = \dfrac{0.168}{0.0168} = 10$  [1]

8 Moles $HCl = 0.550 \times 0.0248 = 1.364 \times 10^{-2}$  [1]

Moles $M_2CO_3 = 1.364 \times 10^{-2} = 6.82 \times 10^{-3}$  [1]

$M_r$ $M_2CO_3 = \dfrac{0.723}{6.82 \times 10^{-3}} = 106$  [1]

$A_r (M_2) = 106 - 60 = 46$ therefore $A_r$ M $= 23$

Metal is sodium  [1]

9 Moles $SO_2 = \dfrac{1000}{64.1} = 15.6$  [1]

(Ratio of $SO_2:SO_3$ is 2:2) moles $SO_3 = 15.6$

Theoretical yield of $SO_3 = 15.6 \times 80.1 = 1250$ g  [1]

% yield $= \dfrac{1225}{1250} \times 100 = 98.0\%$  [1]

10 (a) Initial moles $HCl = 0.650 \times 0.040 = 0.0260$  [1]

Moles unreacted HCl = moles NaOH
$= 0.325 \times 0.0252 = 8.19 \times 10^{-3}$  [1]

Moles HCl used up in reaction =
$0.0260 - 8.19 \times 10^{-3} = 0.0178$  [1]

(b) Moles $CaCO_3 = \dfrac{0.0178}{2} = 8.90 \times 10^{-3}$  [1]

Mass $CaCO_3 = 8.90 \times 10^{-3} \times 100.1 = 0.891$ g  [1]

% age $= \dfrac{0.891}{0.920} \times 100 = 96.8\%$  [1]

## 1.4

1 Transfer of electrons / correct electronic structure of ions  [1]

charges on ions  [1]

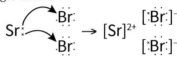

2 (a) $H^{\delta+} - Br^{\delta-}$    $Cl–Cl$    $S^{\delta+} - O^{\delta-}$  [2]

(1 mark if 2 out of 3 correct)

(b) Difference in electronegativity between Ga and Cl is 1.4 so it's polar covalent  [1]

Difference in electronegativity between Ga and F is 2.4 so it's ionic  [1]

3 (a) $O^{\delta+} - F^{\delta-}$  [1]

Fluorine is more electronegative than oxygen  [1]

(b)

$$F \quad O \quad F$$

4 Valence electrons shared to form an electron pair bond.  [1]

The electrons in the pair repel each other  [1]

this is overcome by their attractions to both nuclei  [1]

5 (a) Methane has four equal bond pairs and no lone pairs  [1]

The electron pairs arrange themselves around the central atom as far as possible from each other so that the repulsion between them is at a minimum  [1]

(b) Water has 2 bond pairs and two lone pairs of electrons  [1]

The two lone pairs on the oxygen repel the two bond pairs more strongly than they repel one another so that the bond angle is reduced to 104.5°  [1]

6 $H_2O$    $PH_3$    $SF_6$

7 HCl, HBr and HI have van der Waals forces between the molecules  [1]

The boiling temperature of HI > HBr> HCl since the number of electrons involved in induced dipole interactions also increases  [1]

HF has hydrogen bonds between molecules  [1]

Hydrogen bonds are stronger than van der Waals forces (so more energy is needed to break the bond)  [1]

## 1.5

1

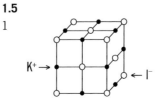

2   In iodine although atoms are covalently bonded in pairs to form diatomic molecules. These molecules are held together by weak van der Waals forces   [1]

Graphite consists of layers where each carbon is joined to three others by strong covalent bonds   [1]

3   Metallic bonding   [1]

Metals consist of a regular arrangement of metal cations (a lattice) surrounded by a 'sea' of delocalised electrons   [1]

The delocalised electrons can carry a current because, when a potential difference is applied across the ends of a metal, they will be attracted to and move towards the positive terminal of the cell   [1]

4   Diamond is a giant covalent structure with each carbon atom covalently bonded to four others in a tetrahedral shape   [1]

Metals consist of a regular arrangement of metal cations (a lattice) surrounded by a 'sea' of delocalised electrons   [1]

There are electrostatic forces of attraction between the nucleus of the cations and the delocalised electrons   [1]

Both have high melting temperatures / are hard   [1]

The energy needed to break the strong covalent bonds or strong metallic bond is very high   [1]

**1.6**

1   Lithium's outer electron is in an s orbital   [1]

2   (a)  Cl, Ar   [1]

(b)  Silicon contributes four electrons to form four strong covalent bonds that are arranged in a tetrahedral shape   [1]

It has a giant covalent structure   [1]

(c)  Argon does not form molecules but exists as separate atoms   [1]

The atoms are held together by very weak induced dipole – induced dipole forces   [1]

(d)  There is an increase because the metallic bonding gets stronger   [1]

The metal ions have a greater charge and there is an increased number of delocalised electrons   [1]

(e)  Chlorine has a greater number of electrons than fluorine   [1]

Therefore there is an increase in the induced dipole – induced dipole intermolecular forces holding the diatomic molecule together   [1]

3   The lead, Pb, has been reduced, oxidation number decreases from +4 to +2   [1]

The antimony, Sb, has been oxidised, oxidation number increases from 0 to +3   [1]

4   (a)

|  | barium chloride | magnesium chloride | potassium carbonate | sodium sulfate |
|---|---|---|---|---|
| barium chloride | ■ | no visible change | white precipitate | white precipitate |
| magnesium chloride |  | ■ | white precipitate | no visible change |
| potassium carbonate |  |  | ■ | no visible change |
| sodium sulfate |  |  |  | ■ |

1 mark for each correct observation

(b)  Barium chloride – apple green flame, magnesium chloride – no change, potassium carbonate – lilac flame, sodium sulfate – orange / yellow flame   [2]

(Any 2 or 3 correct 1 mark)

(c)  Barium carbonate   [1]

$Ba^{2+}(aq) + CO_3^{2-}(aq) \longrightarrow BaCO_3(s)$   [1]

(d)  A white precipitate with the magnesium chloride solution. No visible change with the other three solutions   [2]

(Any 2 or 3 correct 1 mark)

5   (a)  A clean metal wire (or splint) is moisturised with hydrochloric acid (or water), dipped in the compound   [1]

and held in a non-luminous Bunsen flame   [1]

Crimson flame seen   [1]

(b)  (A few drops of nitric acid are added) then add silver nitrate solution   [1]

White precipitate seen   [1]

6   (a)  Colour of green gas would disappear / bright yellow-orange flame   [1]

White solid forms   [1]

(b)  (i)  $2Fe + 3Cl_2 \longrightarrow 2FeCl_3$

(ii)  Iodine is not a strong enough oxidising agent (to oxidise iron to iron(III))

**1.7**

1   (a)  If a system at equilibrium is subjected to a change, the equilibrium tends to shift so as to minimise the effect of the shift

(b)  (i)  Yield of hydrogen chloride decreases   [1]

The system opposes the change by favouring the endothermic direction (so the equilibrium moves to the left)   [1]

(ii)  No change in yield   [1]

There are the same number of moles of gas on each side of the equation   [1]

2   (a)

| Change | Effect, if any, on position of equilibrium | Effect, if any, on value of $K_c$ |
|---|---|---|
| Increase in pressure | Shift to the right | None |
| Drop in temperature | Shift to the right | Increase |

(b)  Low temperature   [1]

Low temperature favours the exothermic direction of a reaction and so a lower temperature will shift equilibrium to the right   [1]

Low pressure   [1]

There are more gaseous moles on the right-hand side so a lower pressure will shift equilibrium to the right   [1]

3   (a)  (Since $H^+(aq)$ ions are on the right-hand side) the concentration of $H^+(aq)$ ions would increase   [1]

The system will try to minimise the effect by decreasing the concentration of chlorine and shifting equilibrium to the right   [1]

(b) A strong acid is one that fully dissociates (in aqueous solution) [1]

A weak acid is one that partially dissociates (in aqueous solution) [1]

(c) $[H^+] = 10^{-4.3}$ [1]

$[H^+] = 5.0 \times 10^{-5}$ mol dm$^{-3}$ [1]

4 (a) $K_c = \dfrac{[CH_3COOC_2H_5][H_2O]}{[CH_3COOH][C_2H_5OH]}$ [1]

no unit [1]

(b) $[C_2H_5OH] = \dfrac{(9.13 \times 10^{-2})^2}{4.07 \times 1.09 \times 10^{-1}}$ [1]

$[C_2H_5OH] = 1.88 \times 10^{-2}$ [1]

5 (a) Moles $CuSO_4.5H_2O = 0.250 \times 0.250 = 0.0625$ [1]

Mass $CuSO_4.5H_2O = 0.0625 \times 249.7 = 15.6$ g [1]

(b) Transfer solid into a beaker and add water [1]

Stir until all the solid dissolves [1]

Add solution to a 250 cm$^3$ volumetric/graduated flask [1]

Using a funnel for transfer / Washing weighing bottle and beaker [1]

Make up to the mark and invert (shake well) to mix [1]

6 (a) Identification of 23.45 cm$^3$ as anomalous result [1]

Mean titre = 22.98 cm$^3$ [1]

(Accept 23.00 cm$^3$)

(b) Add acid into burette using a funnel [1]

Making sure that the jet is filled [1]

Add (a few drops of) indicator to the sodium hydroxide (in the conical flask) [1]

Titrate until the indicator just turns colour [1]

Shake/swirl the conical flask (during titration) [1]

Record burette reading at beginning and end of titration [1]

(c) Less accurate result since titre would be less than 5.00 cm$^3$ / very small [1]

Percentage error is greater in measuring smaller volume [1]

# Unit 2

## 2.1

1 (a) The enthalpy change when one mole of a substance [1] is completely combusted in oxygen under standard conditions [1]

(b) $C_8H_{18}(l) + 12\tfrac{1}{2}O_2(g) \longrightarrow 8CO_2(g) + 9H_2O(g)$ [1]

(c) $\Delta H = (4 \times -394) + (5 \times -286) - (-2878)$ [1]

$\Delta H = -128$ kJ mol$^{-1}$ [1]

(d) (i) $Cl_2(g) + \tfrac{1}{2}O_2(g) \longrightarrow Cl_2O(g)$ [1]

(ii) $76.2 = 242 + 248 - 2(Cl - O)$ [1]

$2(Cl - O) = 413.8$

$Cl - O = 206.9$ kJ mol$^{-1} = 207$ kJ mol$^{-1}$ [1]

2 $q = 75 \times 4.18 \times 4.5$ [1]

$q = 1411$ J [1]

Moles of water formed $= 0.5 \times 0.05 = 0.025$ [1]

$\Delta H = \dfrac{-1411}{0.025}$ [1]

$\Delta H = -56.4$ kJ mol$^{-1}$ [1]

3 (a) The total enthalpy change for a reaction is independent of the route taken from the reactants to the products [1]

(b) (i) $\Delta H = (6 \times -394) + (6 \times -286) - (-1271)$ [1]

$\Delta H = -2809$ kJ mol$^{-1}$ [1]

(ii) Oxygen gas is an element in its standard state [1]

4 (a) (i) $q = 200 \times 4.18 \times 28.2$ [1]

$q = 23575$ J [1]

$-3280 = \dfrac{23.575}{n}$ [1]

$n = 7.19 \times 10^{-3}$ [1]

Mass heptane $= 7.19 \times 10^{-3} \times 100.16 = 0.720$ g [1]

(ii) Heat loss to the environment / incomplete combustion [1]

Increase insulation [1]

(b) $\Delta_c H$ heptane $= -4817$ kJ mol$^{-1}$ [1]

Difference between nonane and decane is 654 kJ mol$^{-1}$ so difference between nonane and heptane (or pentane and heptane) is 1308 kJ mol$^{-1}$ [1]

(c) Energy per gram pentane $= \dfrac{3509}{72.12} = 48.66$ kJ g$^{-1}$ [1]

Energy per gram decane $= \dfrac{6779}{142.22} = 47.67$ kJ g$^{-1}$ [1]

(d) (i) Bonds broken $= (6 \times C-C) + (16 \times C-H) + (9 \times O = O)$ [1]

Bonds formed $= (14 \times C=O) + (16 \times O-H)$ [1]

$\Delta H = (6 \times 348) + (16 \times 412) + (9 \times 496)$
$- ((14 \times 743) + (16 \times 463))$ [1]

$\Delta H = -4666$ kJ mol$^{-1}$ [1]

(ii) Average bond enthalpies were used not actual ones [1]

## 2.2

1 (a) Measure the volume of carbon dioxide produced (using a gas syringe) at timed intervals [1]

Measure the mass of carbon dioxide lost (using a weighing balance) at timed intervals [1]

(b) Crush them into a powder / increase the surface area / stir them continuously [1]

2  (a)  (i)

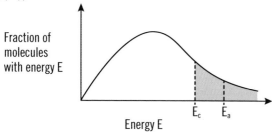

Fraction of molecules with energy E

Energy E

$E_c$  $E_a$

[1]

A greater proportion of colliding molecules will achieve the minimum energy needed to react  [1]

(ii)

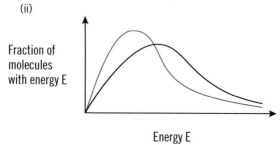

Fraction of molecules with energy E

Energy E

[1]

(b)  (i)  Rate = $\dfrac{0.258}{60}$ = 4.3 × 10⁻³ [1]  mol dm⁻³ s⁻¹  [1]

(ii)  Initial rate would be faster  [1]

Concentration of reactants decreases  [1]

(iii)

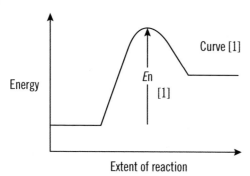

Curve [1]

Energy

$En$

[1]

Extent of reaction

3  (a)  Rate increases  [1]

Increase in kinetic energy of molecules  [1]

More molecules will have required activation energy  [1]

(b)  Rate increases  [1]

More molecules in given volume  [1]

Greater chance for effective collisions  [1]

4  Lower temperatures can be used  [1]

Energy costs saved  [1]

Less (fossil) fuels burned to provide energy so less impact on environment  [1]

More products can be made in a given time (so more profit)  [1]

(Accept any three)

5  (a)  At least 5 points plotted correctly  [1]

Appropriate straight line drawn  [1]

Axes correctly labelled and suitable scale chosen  [1]

(b)  Suitable tangent drawn  [1]

Rate = 1.1 × 10⁻⁴ ± 0.1 × 10⁻⁴  [1]

Unit = mol dm⁻³ s⁻¹  [1]

(c)  Colorimetry method  [1]

Calibrate colorimeter with iodine solution of known concentration  [1]

Measure light passing through to determine concentration at intervals  [1]

(d)  Concentration of hydrogen peroxide is directly proportional to the rate / doubling the concentration of hydrogen peroxide doubles the rate  [1]

Concentration of iodide ions is directly proportional to the rate / doubling the concentration of iodide ions doubles the rate  [1]

**2.3**

1  (a)  $C_4H_{10}(g) + 6\frac{1}{2}O_2(g) \longrightarrow 4CO_2(g) + 5H_2O(g)$  [1]

(b)  (Energy from propane) −49.9 kJ g⁻¹  [1]

(Energy from butane) −48.4 kJ g⁻¹  [1]

(c)  (i)  ($CO_2$ from propane) 2.99 g  [1]

(ii)  ($CO_2$ from butane) 3.03 g  [1]

(d)  Propane since it produces less $CO_2$  [1]

per kJ of energy produced  [1]

2  Ethane since it produces more energy per mole than bioethanol  [1]

However, bioethanol is a renewable fuel / carbon neutral fuel since it is made from sugar cane  [1]

Ethane is a non–renewable fuel so the initial conclusion should be reversed  [1]

3  (a)  Advantage: doesn't produce harmful emissions / renewable / produced from water which is readily available  [1]

Disadvantage: highly flammable / difficult to store / fossil fuels are needed to produce hydrogen from water  [1]

(b)  Advantage: carbon neutral / renewable  [1]

Disadvantage: land used to produce bioethanol cannot be used to grow food / growing crops for producing bioethanol needs large quantities of water (and fertilisers) [1]

4  (a)  More burning of fossil fuels / more industries / more transportation / deforestation  [2]

(Accept any two)

(b)  % increase = $\dfrac{407 - 368}{368}$ × 100 = 10.6 %  [1]

(c)  Not enough data about individual years / Level of $CO_2$ could have decreased during some years  [1]

Smaller gap between 1980 and 2000 so increase in carbon dioxide levels greater (during this period)  [1]

**2.4**

1  (a)  % oxygen = 100 − 55.0 − 9.10 = 35.9  [1]

C  :  H  : O

$\underline{55.0} : \underline{9.10} : \underline{35.9}$

12  1.01  16

4.58 : 9.01 : 2.24  [1]

2.04 : 4.02 : 1

Empirical formula = $C_2H_4O$  [1]

(b) Mass of empirical formula = 44

Therefore, molecular formula = $2 \times C_2H_4O = C_4H_8O_2$ [1]

(c) Carboxylic acid [1]

(d) Butanoic acid [1]

2-methylpropanoic acid [1]

(e) Butanoic acid since it has a longer alkyl chain than 2-methylpropanoic acid. [1]

Longer alkyl chains have larger van der Waals forces between them and so more energy will be needed to separate these molecules (compared to separating 2-methylpropanoic acid molecules) [1]

2 (a)

HO—OH with OH [1]

(b) $C_3H_8O_3$ [1]

(c) Alcohol [1]

(d)

[1]

(e) Halogenoalkanes [1]

(f) Propane-1,2,3-triol because hydrogen bonds can form between the molecules. [1]

1,2,3-trichloropropane only has van der Waals forces between molecules [1]

Hydrogen bonds are stronger, so more energy is needed to separate the propane-1,2,3-triol molecules [1]

3 (a) 2-bromo-1-chloropropane [1]

(b)

Br—Cl [1]

(c)

[1]

(d) $C_3H_6BrCl$ [1]

(e) Halogenoalkanes [1]

(f) 1–bromo-1–chloropropane [1]

1–bromo-2–chloropropane [1]

1–bromo-3–chloropropane [1]

2–bromo-2–chloropropane [1]

## 2.5

1 (a) Hydrocarbons are compounds made up of hydrogen and carbon only [1]

(b) (i) $C_nH_{2n+2}$ [1]

(ii) $C_nH_{2n}$ [1]

(c) Alkenes have double carbon–carbon bonds whereas alkanes have only single carbon–carbon bonds [1]

(d) σ-bonds and π-bonds [1]

σ-bonds are made by the head on overlap of two s-orbitals or by an s and a p-orbital [1]

π-bonds are made by the sideways overlap of two p-orbitals [1]

(e) π-bonds are regions of high electron density and are weaker than σ-bonds (and so are more easily broken) [1]

This leads to the higher reactivity of alkenes compared to alkanes because alkanes do not have π-bonds [1]

2 (a) $C_6H_{14} + 9\frac{1}{2}O_2 \longrightarrow 6CO_2 + 7H_2O$ [1]

(b) $C_xH_y + excess O_2 \longrightarrow xCO_2 + y/2H_2O$

$x = 6.1/44 = 0.139$

$y = 2 \times (3.0/18.02) = 0.333$ [1]

Ratio is 1 to 2.4 [1]

So simplest integer ratio is 5 to 12 and so the empirical formula = $C_5H_{12}$ [1]

(c) Carbon monoxide gas would be made instead [1]

Carbon monoxide is poisonous/toxic because it combines with haemoglobin in the blood thus reducing the ability to carry oxygen around the body [1]

(d) Answers can be any two of the following. One mark for issue and one mark for disadvantage: [4]

Fossil fuels are non-renewable – this means that they will ultimately run out and so alternative energy sources will have to be used

Many fossil fuels contain sulfur and when they burn sulfur dioxide is made which leads to the formation of acid rain. Acid rains can cause serious damage to buildings, trees, etc. (Alternatively, sulfur dioxide is a health problem for people with breathing difficulties)

Nitrogen oxides can be produced in combustion engines (due to the high temperature at which they operate) and this leads to the formation of acid rain. Acid rain can cause serious damage to buildings, trees, etc. (Alternatively, nitrogen oxides can be a health problem for people with breathing difficulties)

Combustion of fossil fuels creates carbon dioxide which is a greenhouse gas. The excess production of greenhouse gases is causing global warming and climate change which has serious consequences such as rising sea levels

3 (a) But-1-ene and but-2-ene [1]

(b) But-2-ene exists as E–Z isomers [1]

Each end of the carbon-carbon double bond has two different groups attached to it [1]

(c) (i) Electrophilic addition [1]

(ii) 2-bromopropane [1]

(iii) Diagram should show:

Correct dipole on $^{\delta+}H–Br^{\delta-}$ [1]

Two curly arrows – one from double bond to $^{\delta+}H$ and second from H-Br bond to the $^{\delta-}Br$ [1]

Secondary carbocation intermediate and curly arrow from lone pair on $Br^-$ to the positive carbon [1]

(iv) 2-bromopropane is formed from the more stable secondary carbocation (as opposed to the alternative, less stable primary carbocation) [1]

## 2.6

1 (a) Nucleophilic substitution. [1]

(b) A species with a lone pair of electrons, which can be donated to an electron deficient species. [1]

(c) Carbon–halogen bond is polarised so there is a $C^{\delta+}$ which is susceptible to attack by nucleophiles. [1]

This arises because halogens are more electronegative than carbon. [1]

(d) Diagram for mechanism needs to include:
- Polarisation of C–Cl bond [1]
- Lone pair on attacking OH⁻ and Cl⁻ produced [1]
- Curly arrow from lone pair on OH⁻ and curly arrow showing breaking of C–Cl bond. [1]

(e)

OH [1]

(f) 2-chloro-2-methylbutane or 2-bromo-2-methylbutane [1]

(g) (Sodium hydroxide dissolved in) ethanol. [1]

Elimination. [1]

(h) But-1-ene [1]

(E/Z-) but-2-ene. [1]

2 (a) Suitable overall diagram similar to this:

Water out
Condenser
Water in
Flask
Reagents
Heat [1]

vertical condenser (water in at bottom and out at top) [1]

heat [1]

All volatile compounds are condensed and returned to the flask [1]

so prolonged heating is possible without any loss of material [1]

(b) Distillation [1]

1-Bromopropane has a lower boiling point that propan-1-ol because there are only van der Waals forces between its molecules [1]

There are hydrogen bonds between the molecules of propan-1-ol, and hydrogen bonds are stronger than van der Waals forces [1]

(c) Take a (small) sample of the reaction mixture out of the flask and add (excess) nitric acid [1]

Then add a solution of silver nitrate [1]

A cream precipitate will confirm the presence of bromide ions [1]

(d) The reaction would be faster with 1-iodopropane [1]

The C–I bond is weaker/has a lower bond enthalpy than the C–Cl bond [1]

3 (a) UV light breaks C–Cl bonds (in halogenoalkanes/CFCs) producing chlorine radicals [1]

These radicals react with ozone and thus the ozone layer is damaged [1]

The ozone layer protects us from UV radiation [1]

UV radiation causes skin cancer [1]

(b) Scientists have developed alternatives called HFCs [1]

which contain only C–H and C–F bonds. C–F bonds are stronger than C–Cl bonds and so are not broken by UV light [1]

(c) $E = hc/\lambda$   so $\lambda = hc/E$ [1]

Energy per atom = $413 \times 1000 / 6.02 \times 10^{23}$ = $6.86 \times 10^{-19}$ J [1]

$\lambda = 6.63 \times 10^{-34} \times 3.00 \times 10^8 / 6.86 \times 10^{-19}$

= $2.899 \times 10^{-7}$ m = 290nm [1]

C–H bonds will be broken by light of wavelength 290 nm and not 400 nm (which does not have sufficient energy to break a C–H bond) [1]

**2.7**

1 (a) Reflux each sample with acidified potassium dichromate [1]

The one that does not give a colour change (orange to green) is the tertiary alcohol [1]

Primary and secondary alcohols will be oxidised by potassium dichromate but tertiary alcohols cannot be oxidised [1]

Isolate the remaining two products and test with sodium hydrogencarbonate solution [1]

The one which effervesces is the primary alcohol, the one which does not effervesce is the secondary alcohol [1]

On refluxing (complete oxidation will occur), the primary alcohol will be oxidised to a carboxylic acid (and the secondary alcohol to a ketone). Carboxylic acids react with sodium hydrogencarbonate solution to release carbon dioxide gas, ketones do not [1]

(b) Primary alcohol could be butan-1-ol or 2-methylpropan-1-ol [1]

Secondary alcohol is butan-2-ol [1]

Tertiary alcohol is 2-methylpropan-2-ol [1]

(c) Boiling point [1]

All the alcohols will have hydrogen bonding between the molecules, but the structures with the least branched alkyl chain will have the highest boiling point [1]

This part of the molecule will have the greatest van der Waals forces between the structures [1]

2 (a) For each reaction: reagents and conditions [1], correct overall equation [1], unambiguous organic product [1]

**Dehydration**

Reagent(s): Conc. $H_2SO_4$ / conc. $H_3PO_4$ / aluminium oxide.

Conditions: warm/heat

$$CH_3CH(OH)CH_3 \longrightarrow CH_3CH=CH_2 + H_2O$$

**Oxidation**

Reagent(s): Acidified potassium dichromate / acidified potassium manganate(VII)

Conditions: warm/heat

$$CH_3CH(OH)CH_3 + [O] \longrightarrow CH_3COCH_3 + H_2O$$

**Esterification**

Reagent(s): Conc. $H_2SO_4$, and a carboxylic acid e.g. ethanoic acid.

Conditions: warm/heat

$$CH_3CH(OH)CH_3 + CH_3COOH \longrightarrow CH_3COOCH(CH_3)_2 + H_2O \text{ (arrow can be reversible)}$$

(b)

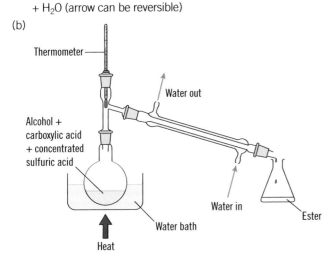

Suitable overall diagram [1]

Thermometer adjacent to the outlet leading to the condenser [1]

Water in at bottom and out at top [1]

This technique works because esters have lower boiling points than both the acid and the alcohol they are made from [1]

Acids and alcohols have hydrogen bonds between the molecules, and these are stronger than the van der Waals forces between the ester molecules [1]

3 (a) $C_2H_4 + H_2O \longrightarrow CH_3CH_2OH$ [1]

$C_6H_{12}O_6 \longrightarrow 2CH_3CH_2OH + 2CO_2$ [1]

(b) Essential conditions – any 2: [2]

Needs yeast/enzyme/zymase

Temperature should be between 25 and 40°C

Process carried out in the absence of oxygen

(c) Advantages – any 2 sensible answers such as: [2]

The raw material is renewable

Only low temperatures and low pressures are required

It can be a carbon-neutral process

Disadvantages – any 2 sensible answers such as: [2]

Large quantities of land are needed to grow crops for biofuel

The land needed could be used for growing crops for food instead

Growing crops for biofuels needs large quantities of water (and fertilisers)

**2.8**

1 The molecular ion is a positive ion formed from the whole molecule, with its mass corresponding to the molecular mass ($M_r$) of the compound [1]

It is usually the peak furthest to the right on a mass spectrum [1]

2 $CH_3^+$ 15      $CH_3CO^+$ 43      $CO^+$ 28

$CH_3COCH_3^+$ 58 (molecular ion)

(All correct, 2 marks; Any 2 or 3 correct, 1 mark)

3 C=O 1650–1750 cm$^{-1}$. C–H 2800–3100 cm$^{-1}$

(both correct for 1 mark)

4 (a) The main difference would be that the IR spectrum for propan-2-ol would have no peak present at 1650–1750 cm$^{-1}$. but would have a broad peak at 3200–3550 cm$^{-1}$ which would be absent from the propanone spectrum [1]

(b) As the propan-2-ol is oxidised into propanone the broad absorption at 3200–3550 cm$^{-1}$ for the –OH group would no longer be present [1]

The sharp C=O peak at 1650–1750 cm$^{-1}$ would appear, providing evidence that the secondary alcohol has been oxidised into a ketone [1]

5 –COOH [1]

There is a characteristic C=O absorption at 1650–1750 cm$^{-1}$ as well as a very broad absorption at 2500–3200 cm$^{-1}$ for the –OH group [1]

6 2 peaks, one at δ 190 to 220 ppm for the C=O and another at 20 to 50 ppm for the identical $CH_3$ groups. [1]

7 3 peaks, one at δ 160 to 185 ppm for the C=O on the ester functional group [1] another at 20 to 50 ppm for the $CH_3$ group bonded to the ester C=O (part of the ethanoate group), [1] and one at 50 to 90 ppm for the $CH_3$ group bonded to the oxygen in the ester functional group [1]

8 One at δ 0.9 ppm for the $CH_3$ group with a peak area of 3

another at 2 to 3 ppm with a peak area of 2 for the $CH_2$ group (bonded to the COOH) [1]

and one at 11 ppm with a peak area of 1 for the (acidic) COOH proton [1]

9 Propanal [1]

# Answers to questions

## Answers to questions on pages 6-7

| AO | Question | Mark scheme |
|---|---|---|
| 1 | Demonstrate knowledge | Ability to attract electrons in a covalent bond |
| 1 | Demonstrate understanding of scientific ideas | At higher concentrations, more particles in a given volume / particles closer together (1)<br>More frequent collisions / more chance of collisions (1) |
| 2 | In a theoretical context | precipitate of barium sulfate forms with barium chloride (1)<br>none with magnesium chloride as magnesium sulfate is soluble (1)<br>$Ba^{2+}(aq) + SO_4^{2-}(aq) \rightarrow BaSO_4(s)$ or $BaCl_2(aq) + H_2SO_4(aq) \rightarrow BaSO_4(s) + 2HCl(aq)$ (1) |
| 2 | In a practical context | Award (1) each for correct reagent and observation<br>Award (1) for correct √/× for all three compounds for first test<br>Award (1) for correct √/× for all three compounds for second test<br>Award (1) for tests that enable all three compounds to be identified<br><br>(see table below) |

| Reagent(s) | Observation expected for positive result | butan-2-ol | 2-methylpropanoic acid | 3-hydroxybutanoic acid |
|---|---|---|---|---|
| $Na_2CO_3$ / Mg | Effervescence / fizzing / bubbles | × | √ | × |
| acidified potassium dichromate / acidified potassium manganate(VII) / iodine and sodium hydroxide | orange to green / purple to colourless / yellow precipitate | √ | × | √ |

| AO | Question | Mark scheme |
|---|---|---|
| 2 | When handling qualitative data | Displayed formulae of butan-1-ol and butan-2-ol for (1) each<br><br>H H H H<br>  &#124; &#124; &#124; &#124;<br>H−C−C−C−C−OH  butan-1-ol<br>  &#124; &#124; &#124; &#124;<br>  H H H H<br><br>H H H H<br>  &#124; &#124; &#124; &#124;<br>H−C−C−C−C−H  butan-2-ol<br>  &#124; &#124; &#124; &#124;<br>  H H OH H |
| 2 | When handling quantitative data | $V = \dfrac{nRT}{P}$     (1)<br>$V = \dfrac{2.54 \times 10^{-3} \times 8.31 \times 393}{101000}$     (1)<br>$V = 8.21 \times 10^{-5}$ m$^3$     (1)<br>$V = 82.1$ cm$^3$     (1) |
| 3 | Make judgements and reach conclusions | $[ClF_2]^+$ has 2 bond pairs and 2 lone pairs (1)<br>$[ClF_2]^-$ has 2 bond pairs and 3 lone pairs (1)<br>Different number of electron pairs so different shapes (and student is correct) (1) |
| 3 | Develop and refine practical design and procedures | Any two of the following<br>• Record the steady temperature of the water before adding ammonium nitrate (1)<br>• $\Delta T$ can be calculated accurately (1)<br>  or<br>• Record the temperature at timed intervals after adding ammonium nitrate (1)<br>• Graph can be extrapolated (1)<br>  or<br>• Place lid on cup / insulate cup (1)<br>• Prevents heat loss (1)<br>  or<br>• Crush the ammonium nitrate (1)<br>• Solid dissolves more quickly (1) |

# Exam practice answers

## Unit 1

### Answering examination questions Unit 1

(a) Energy required to remove an electron from each atom in 1 mole of element in the gaseous state. (Accept appropriate equation) (1)

(b) Higher because the outer electron is closer to the nucleus (1) and there is less shielding. (1)

(c) Energy per atom = $736 \times 1000/6.02 \times 10^{23} = 1.22 \times 10^{-18}$ J (1)

Frequency = $1.22 \times 10^{-18}/6.63 \times 10^{-34} = 1.84 \times 10^{15}$ Hz (1)

Wavelength = $3.00 \times 10^8/1.84 \times 10^{15} = 1.63 \times 10^{-7}$ m (1)

Wavelength = 163 nm (1) (Answer must be to 3 sig figs)

### Exam practice questions

The following mark scheme is based on that provided by WJEC, but WJEC bears no responsibility for the answers provided within this publication.

| Question | | | Marks available | Marking details | | | | | |
|---|---|---|---|---|---|---|---|---|---|
| | | | | A01 | A02 | A03 | Total | Maths | Prac |
| 1 | (a) | | Diagram or description of face centred cubic for NaCl (1) <br> Diagram or description of body centred cubic for CsCl (1) <br> Co-ordination 6:6 in NaCl and 8:8 in CsCl (1) <br> $Cs^+$ bigger than $Na^+$ so more $Cl^-$ can fit round the metal ion (1) | 4 | | | 4 | | |
| | (b) | | Any **four** of following – 4 max <br> (For a liquid to boil) intermolecular bonds/forces must be broken (1) <br> $H_2O$ held together by hydrogen bonds (1) <br> S less electronegative (1) <br> $H_2S$ held together by van der Waals forces (1) <br> Van der Waals forces are weaker than hydrogen bonds (1) | | 4 | | 4 | | |
| | (c) | | $PCl_6^-$ is octahedral and $PCl_4^+$ is tetrahedral – credit diagram (1) <br> Bonds angles of 90° and 109–110° (1) <br> (Due to presence of) 6 bond pairs and 4 bond pairs (1) <br> Arranged to give minimum repulsion/ maximum separation (1) | 1 <br> 1 <br> 1 | 1 | | 4 | | |
| | | | **Question 1 total** | 7 | 5 | 0 | 12 | 0 | 0 |
| 2 | (a) | (i) | Energy needed to remove one electron from (every atom) in 1 mol of element in the gaseous state (1) under standard conditions (1) | 2 | | | 2 | | |
| | | (ii) I | (For ionisation to take place) an electron must overcome the attraction between it and the nucleus (1) this increases for successive ionisations as each electron is being removed from an increasingly positive species / ratio of protons : electrons goes up (1) | 2 | | | 2 | | |
| | | II | X is in Group II – some attempt at explanation required (1) <br> big jump after removal of two electrons / new shell broken into after removal of two electrons (1) | | | 2 | 2 | | |
| | (b) | (i) | **Indicative content** <br> • for ionisation energy determination transitions to $n = 1$ must be used / Lyman series must be used (**statement 1**) <br> • lines in visible spectrum comes from transitions involving $n = 2$ / Balmer series <br> • electrical energy excites electrons <br> • electrons fall back and give off energy <br> • lines formed since only some energies are allowed <br> • lines are closer together at higher energies <br> • since energy levels get closer <br> • convergence limit is when electron removed / ionisation takes place <br> • use $E = hf$ <br> **5–6 marks** <br> Recognition that Lyman series required, clear understanding of electron transition and convergence of lines/energy levels <br> *The candidate constructs a relevant, coherent and logically structured account including all key elements of the indicative content. A sustained and substantiated line of reasoning is evident and scientific conventions and vocabulary are used accurately throughout.* | 5 | | 1 | 6 | | |

Continued ▶

| Question | | | Marks available | A01 | A02 | A03 | Total | Maths | Prac |
|---|---|---|---|---|---|---|---|---|---|
| | | | **3–4 marks**<br>Basic understanding of electron transition and convergence of lines/energy levels<br>*The candidate constructs a coherent account including many of the key elements of the indicative content. Some reasoning is evident in the linking of key points and use of scientific conventions and vocabulary is generally sound.*<br>**1–2 marks**<br>Some knowledge of electron transition between energy levels<br>*The candidate attempts to link at least two relevant points from the indicative material. Coherence is limited by omission and/or inclusion of irrelevant materials. There is some evidence of appropriate use of scientific conventions and vocabulary.*<br>**0 marks**<br>*The candidate does not make any attempt or give an answer worthy of credit.* | | 3 | | 3 | 3 | |
| | (ii) | | $\Delta E = 6.63 \times 10^{-34} \times 3.28 \times 10^{15} (= 2.175 \times 10^{-18}$ J) (1)<br>$6.63 \times 10^{-34} \times 3.28 \times 10^{15} \times 6.02 \times 10^{23}$ (1)<br>1310 (1) answer **must** be given to 3 sig figs | | 3 | | 3 | 3 | |
| | | | **Question 2 total** | 9 | 3 | 3 | 15 | 3 | 0 |
| 3 | (a) | (i) | mass water = 0.274 (g)<br>mass anhydrous barium chloride = 1.645 (g) (1) **both** required<br>correct $M_r$ values for water and barium chloride $\Rightarrow$ 18 and 208 (1) **both** required<br>$\dfrac{1.645}{208} : \dfrac{0.274}{18} \Rightarrow 0.0079 : 0.0152$ (1)<br>1 : 1.92 therefore x = 2 (1) | | 3 | 1 | 4 | 3 | 4 |
| | | (ii) | To avoid loss by spitting / fumes / loss of solid do **not** accept 'to avoid loss of water' | 1 | | | 1 | | 1 |
| | | (iii) | Use a greater mass of hydrated solid (1) increases **percentage** accuracy (1)<br>Ensure that all water has been lost (1) heat to constant mass (1)<br>Neutral answer: 'heat for longer or hotter' | | | 4 | 4 | | 4 |
| | | (iv) | $x$ is a whole number so small variation in answer is irrelevant – some comment required | | | 1 | 1 | | 1 |
| | (b) | (i) | Carbonate / $CO_3^{2-}$ do **not** accept sulfate / $SO_4^{2-}$ | | | 1 | 1 | | 1 |
| | | (ii) | Any metal ion apart from Group I | | 1 | | 1 | | 1 |
| | | (iii) | $Ba^{2+}(aq) + CO_3^{2-}(aq) \rightarrow BaCO_3(s)$<br>ecf possible, e.g. if sulfate given in (i) | | 1 | | 1 | | |
| | | | **Question 3 total** | 1 | 5 | 7 | 13 | 3 | 12 |
| 4 | (a) | (i) | (+)5 | | 1 | | 1 | | |
| | | (ii) | $n(NaClO_3) = 0.826$ (1)<br>$n(O_2) = 0.826 \times 1.50 = 1.239$ (1)<br>$V = \dfrac{nRT}{p} = \dfrac{1.239 \times 9.31 \times 600}{1.01 \times 10^5} = 61.2$ dm³ (1) **must** be to 3 sig figs<br>Accept molar volume method<br>$V = \dfrac{1.239 \times 22.4 \times 600}{273} = 61.0$ dm³ (1) **must** be to 3 sig figs | | 2 | 1 | 3 | 3 | |
| | (b) | | $M_r(NaClO) = 74.5$<br>total $M_r$ of reactants = 151.02 (1) **both** needed<br>atom economy = $\dfrac{74.5}{151.02} \times 100 = 49.3$ % (1) | | 2 | | 2 | 1 | |
| | (c) | | $\dfrac{18.8}{23} : \dfrac{29.9}{35.5} : \dfrac{52.2}{16}$<br>0.817 : 0.817 : 3.27 (1)<br>1 : 1 : 4<br>therefore empirical formula is $NaClO_4$ (1) | | 2 | | 2 | 1 | |
| | | | **Question 4 total** | 0 | 7 | 1 | 8 | 5 | 0 |

Continued ▶

| Question | | | Marks available | Marking details | | | | | |
|---|---|---|---|---|---|---|---|---|---|
| | | | | A01 | A02 | A03 | Total | Maths | Prac |
| 5 | (a) | | a reaction in which products react to form reactants (as well as reactants reacting to form products) | 1 | | | 1 | | |
| | (b) | | both lines starting as curves and then being horizontal at same time as original (1)<br>line A starting at 2.5 and finishing at 0.5 (1)<br>line B starting at 0 and finishing at 2.0 (1) | | | 3 | 3 | 2 | |
| | (c) | (i) | $K_c = \dfrac{[CH_3COOC_2H_5][H_2O]}{[CH_3COOH][C_2H_5OH]}$<br>no units (must follow $K_c$) (1) | | 2 | | 2 | 1 | |
| | | (ii) | $\Delta H$ approx $= 0$ (1)<br>explanation in terms of le Chatelier's principle (1) | | 1 | 1 | 2 | | |
| | (d) | | pH $= -\log[H^+]$ (1)<br>$[H^+] = 3.98 \times 10^{-3}$ mol dm$^{-3}$ (1) | 1 | 1 | | 2 | 2 | |
| | (e) | | moles of $CH_3COOH = 2.94/60 = 0.049$ and moles $C_2H_5OH = 0.045$ (1)<br>moles ethanol is limiting factor (1)<br>theoretical yield $CH_3COOC_2H_5 = 0.045 \times 88 = 3.96$ g (1)<br>percentage yield $= 2.73/3.96 \times 100 = 68.9$ % (1) | 1 | 3 | | 4 | 2 | |
| | | | **Question 5 total** | 3 | 7 | 4 | 14 | 7 | 0 |

# Unit 2

## Answering examination questions Unit 2

(a) (i) $C_8H_{18}(l) + 12\frac{1}{2}O_2(g) \longrightarrow 8CO_2(g) + 9H_2O(l)$     (1)

   (ii) The energy required to break bonds is less than the energy released when bonds are formed.     (1)

   (iii) More bonds are broken and made in octane, therefore more energy is released.     (1)

(b) Octane in spirit burner with flame     (1)

  Water in insulated container / calorimeter     (1)

  Suitable overall diagram showing minimum heat loss     (1)

(c) (i) Less accurate because temperature rise / loss in mass is very small     (1)

     therefore greater percentage error     (1)

   (ii) Less accurate because mass of fuel used after boiling did not result in temperature rise     (1)
     therefore calculated value of $\Delta_cH$ too small

## Exam practice questions

The following mark scheme is based on that provided by WJEC, but WJEC bears no
responsibility for the answers provided within this publication.     (1)

| Question | | | Marks available | Marking details | | | | | |
|---|---|---|---|---|---|---|---|---|---|
| | | | | AO1 | AO2 | AO3 | Total | Maths | Prac |
| 1 | (a) | | **Indicative content**<br>• Alkenes react more readily than alkanes<br>• Alkanes react by radical substitution / photohalogenation<br>• Alkanes are unreactive since they contain strong σ-bonds only<br>• Alkenes react by electrophilic addition<br>• Alkenes contain σ-bonds and π-bonds<br>• π-bond is weaker than σ-bond so is easily broken<br>• π-bond gives region of high electron density<br>**5–6 marks**<br>Names both types of reaction and fully explains difference in reactivity.<br>*The candidate constructs a relevant, coherent and logically structured method including all key elements of the indicative content. A sustained and substantiated line of reasoning is evident and scientific conventions and vocabulary are used accurately throughout.*<br>**3–4 marks**<br>Names at least one type of reaction and partially explains difference in reactivity.<br>*The candidate constructs a coherent account including most of the key elements of the indicative content. Some reasoning is evident in the linking of key points and use of scientific conventions and vocabulary are generally sound.*<br>**1–2 marks**<br>Names types of reaction but gives no explanation or simply explains why alkenes are more reactive but does not name reaction types.<br>*The candidate attempts to link at least two relevant points from the indicative content. Coherence is limited by omission and/or inclusion of irrelevant material. There is some evidence of appropriate use of scientific conventions and vocabulary.*<br>**0 marks**<br>*The candidate does not make any attempt or give an answer worthy of credit.* | 5 | | 1 | 6 | | |
| | (b) | | Heat with aqueous sodium hydroxide (1)<br>Add nitric acid and aqueous silver nitrate (1)<br>White precipitate observed (1) | 3 | | | 3 | | |
| | (c) | (i) | H  H  H  H<br>\|   \|   \|   \|<br>C = C – C – C – H<br>\|     \|   \|<br>H     H  H | | 1 | | 1 | | |
| | | (ii) | Dissolved in ethanol (and heated) | 1 | | | 1 | | |
| | (d) | | It has stronger/more Van der Waals forces (1)<br>because it has a larger surface area/hydrocarbon chain (1) | | 2 | | 2 | | |
| | (e) | | Non-flammable (1)<br>Non-toxic (1)<br>(Accept suitable volatility) | 2 | | | 2 | | |
| | (f) | | C      F     O         C     F     O<br>$\frac{14.0}{12} : \frac{44.5}{19} : \frac{41.5}{16}$   $\frac{1.17}{1} : \frac{2.34}{2} : \frac{1.17}{1}$ (1)<br>Empirical formula is $CF_2Cl$ (1) | | 1<br><br>1 | | 1 | | |

Continued ▶

| Question | | | Marks available | AO1 | AO2 | AO3 | Total | Maths | Prac |
|---|---|---|---|---|---|---|---|---|---|
| | (f) | | Mass spectrum shows two chlorines in molecule / shows $M_r$ is 171 (1) | | 1 | | | | |
| | | | Molecular formula is $C_2F_4Cl_2$ (1) | | 1 | | | | |
| | | | Only one carbon environment so formula is | | | | | | |
| | | | $$Cl-\overset{\displaystyle F}{\underset{\displaystyle F}{C}}-\overset{\displaystyle F}{\underset{\displaystyle F}{C}}-Cl$$ | | | 1 | 5 | | |
| | | | **Question 1 total** | 11 | 8 | 1 | 20 | 1 | 0 |
| 2 | (a) | | $\Delta H = \dfrac{-mc\Delta T}{n}$ (1) | 1 | | | | | |
| | | | $\Delta T = 6.8°C$, m = 200 g, n = $2.043 \times 10^{-3}$ (1) | | | | | | |
| | | | $\Delta H = -2782575$ (1) | | 3 | | 4 | 4 | |
| | | | $\Delta H = -2783$ kJ mol$^{-1}$ (1) | | | | | | |
| | (b) | (i) | $\dfrac{0.2}{6.8} \times 100 = 2.9$ | 1 | | | 1 | 1 | 1 |
| | | (ii) | use less water (e.g. 50 cm$^3$) / burn for longer (1) | | | | | | |
| | | | temperature rise higher so percentage error less (1) | | | 2 | 2 | | 2 |
| | (c) | (i) | reduce distance between flame and beaker / protect flame from draughts | | | 1 | 1 | | 1 |
| | | (ii) | incomplete combustion / thermal capacity of beaker | | | 1 | 1 | | 1 |
| | (d) | (i) | $C_5H_{11}OH + 7\frac{1}{2}O_2 \rightarrow 5CO_2 + 6H_2O$ | | 1 | | 1 | | |
| | | (ii) | $\Delta_fH^{\ominus}$ reactants = $-380$ and $\Delta_fH^{\ominus}$ products = $-3686$ (1) | | | | | | |
| | | | $\Delta_cH^{\ominus} = -3686 - (-380) = -3306$ kJ mol$^{-1}$ (1) | | 2 | | 2 | 2 | |
| | | | ecf possible from part (i) | | | | | | |
| | | (iii) | oxygen gas is an element in its standard state | 1 | | | 1 | | |
| | (e) | | reflux both with acidified potassium dichromate (1) | | | | | | |
| | | | colour changes from orange to green with pentan-2-ol (1) | | 3 | | 3 | | |
| | | | no change with 2-methylbutan-2-ol (1) | | | | | | |
| | (f) | (i) | thermometer bulb adjacent to outlet leading to condenser (1) | | | | | | |
| | | | water in through lower tube and out through upper tube (1) | 2 | | | 2 | | 2 |
| | | (ii) | mass of alcohol = $5 \times 0.805 = 4.025$ | | | | | | |
| | | | moles of alcohol = $4.025 \div 88 = 0.0457$ (1) | | | | | | |
| | | | theoretical mass of chloroalkane = $0.0457 \times 106.5 = 4.87$ | | | | | 2 | |
| | | | actual mass chloroalkane = $4.05 \times 0.866 = 3.51$ g (1) | | | 3 | 3 | | |
| | | | percentage yield = $3.51/4.87 \times 100 = 72\%$ therefore student is incorrect (1) | | | | | | |
| | | | **Question 2 total** | 4 | 10 | 7 | 21 | 9 | 7 |
| 3 | (a) | | gas syringe | 1 | | | 1 | | 1 |
| | | | accept collection over water using graduated collection vessel | | | | | | |
| | (b) | | use a weighing balance to measure mass of $CO_2$ / gas lost (1) | | | 1 | | | |
| | | | over time (1) | 1 | | | 2 | | 2 |
| | (c) | | moles of $MgCO_3 = 5.9 \times 10^{-3}$ | | | | | | |
| | | | moles of HCl = $8.0 \times 10^{-3}$ (1) | | | | | 1 | |
| | | | since ratio $MgCO_3$ : HCl is 1:2 only $4.0 \times 10^{-3}$ moles of $MgCO_3$ react therefore carbonate is in excess / he is correct (1) | | | 2 | 2 | | |
| | (d) | | initial rate = 32 (1) (exact value depends on tangent drawn) | | | | | | |
| | | | unit cm$^3$ min$^{-1}$ (1) | | 2 | | 2 | 2 | 2 |
| | (e) | | curve drawn is less steep than the original and falls short of / levels off at 48 cm$^3$ (1) | | | 1 | | | |
| | | | because there is a decrease in concentration so there are fewer successful collisions per unit time (1) | | 2 | | | | |
| | | | volume amount of carbon dioxide has halved since the number of moles of hydrochloric acid has halved (1) | | | | 3 | | |
| | (f) | | surface area of magnesium carbonate | 1 | | | 1 | | 1 |
| | | | **Question 3 total** | 3 | 4 | 4 | 11 | 3 | 6 |

# Glossary

**Acid** is a proton ($H^+$) donor.

**Acid rain** Rain with lower than expected pH.

**Activation energy** The minimum energy required to start a reaction by breaking of bonds.

**Addition reaction** Reaction in which reagents combine to give only one product.

**Alcohol** Homologous series containing –OH as the functional group.

**Aldehyde** Homologous series containing –CHO as the functional group.

**Atom economy** Total $M_r$ of required product x 100% ÷ total $M_r$ of reactants.

**Atomic number (Z)** The number of protons in the nucleus of an atom.

**Atomic orbital** A region in an atom that can hold up to two electrons with opposite spins.

**Average bond enthalpy** The average value of the enthalpy required to break a given type of covalent bond in the molecules of a gaseous species.

**Avogadro constant** Number of atoms per mole.

**Base** is a proton ($H^+$) acceptor.

**Biofuel** A fuel that has been produced using a biological source.

**Bond enthalpy** The enthalpy required to break a covalent X–Y bond into X atoms and Y atoms, all in the gas phase.

**Carbocation** is an ion with a positively charged carbon atom.

**Carbon neutral** A process is carbon neutral if there is no net transfer of carbon dioxide to or from the atmosphere.

**Carbon neutrality** Chemical process that does not lead to an overall increase in carbon dioxide levels.

**Carboxylic acid** Homologous series containing –COOH as the functional group.

**Catalyst** A substance that increases the rate of a chemical reaction without being used up in the process. It increases the rate of reaction by providing an alternative route of lower activation energy.

**CFCs** Halogenoalkanes containing both chlorine and fluorine.

**Chain reaction** A reaction that involves a series of steps and, once started, continues.

**Characteristic absorption** The wavenumber range at which a particular bond absorbs radiation.

**Chemical shift (δ)** The measure of the difference in parts per million (ppm) of the frequency of absorption compared to the absorption of a standard which has a reference peak at 0 ppm.

**Classification of alcohols** Primary, secondary or tertiary – according to structure.

**Complete combustion** Combustion that occurs with excess oxygen.

**Convergence limit** When the spectral lines become so close together they have a continuous band of radiation and separate lines cannot be distinguished.

**Coordinate bond** A covalent bond in which both shared electrons come from one of the atoms.

**Coordination number** The number of atoms or ions immediately surrounding a central atom in a complex or crystal

**Covalent bond** A pair of electrons with opposed spin shared between two atoms with each atom giving one electron.

**Delocalised** An electron that is not attached to a particular atom – it can move around between atoms.

**Displayed formula** Shows all the bonds and atoms in the molecule.

**Dynamic equilibrium** When the forward and reverse reactions occur at the same rate.

**Electronegativity** A measure of the electron-attracting power of an atom in a covalent bond.

**Electronic configuration** The arrangement of electrons in an atom.

**Electrophile** An electron-deficient species that can accept a lone pair of electrons.

**Elimination reaction** A reaction that involves the loss of a small molecule to produce a double bond.

**Empirical formula** The simplest formula showing the simplest whole number ratio of the number of atoms of each element present.

**Endothermic reaction** A reaction that takes in energy from the surroundings, there is a temperature drop and ΔH is positive.

**Enthalpy change (ΔH)** The heat added to a system at constant pressure.

**Enthalpy (H)** The heat content of a system at constant pressure.

**Environment** The nature of the surrounding atoms/groups in a molecule.

**Exothermic reaction** A reaction that releases energy to the surroundings, there is a temperature rise and ΔH is negative.

**E-Z isomerism** Isomerism that occurs in alkenes (and substituted alkenes) due to restricted rotation about the double bond.

**Fermentation** Enzyme-catalysed reaction that converts sugars to ethanol.

**First ionisation energy of an element** The energy required to remove one electron from each atom in one mole of its gaseous atoms.

**Fossil fuel** Fuels derived from organisms that lived long ago.

**Fragmentation** Splitting of molecules, in a mass spectrometer, into smaller parts.

**Functional group** Refers to the atom/group of atoms that gives the compound its characteristic properties.

**Greenhouse gas** A gas that causes an increase in the Earth's temperature.

**Half-life** Time taken for half the atoms in a radioisotope to decay. The time taken for the radioactivity of a radioisotope to fall to half its initial value.

**Halogenation** A reaction with any halogen.

**Halogenoalkane** An alkane in which one or more hydrogen atoms have been replaced by a halogen.

**Hess's law** The total enthalpy change for a reaction is independent of the route taken from the reactants to the products.

**Heterogeneous catalyst** A catalyst in a different physical state from the reactants.

**Heterolytic bond fission** When a bond is broken and one of the bonded atoms receives both electrons from the covalent bond. Ions are formed.

**HFCs** Halogenoalkanes containing fluorine as the only halogen.

**Homogeneous catalyst** A catalyst in the same physical state as the reactants.

**Homologous series** A series of compounds with the same functional group.

**Homolytic bond fission** When a bond is broken and each of the bonded atoms receives one of the bond electrons.

**Hydrocarbon** A compound of carbon and hydrogen only.

**Hydrolysis** A reaction with water to produce a new product.

**Incomplete combustion** Combustion that occurs with insufficient oxygen.

**Initiation** The reaction that starts the process.

**Intermolecular bonding** The weak bonding holding the molecules together. This governs the physical properties of the substance.

**Intramolecular bonding** The strong bonding between the atoms in the molecule. This governs its chemistry.

**Ion** A particle where the number of electrons does not equal the number of protons.

**Ionic bond** A bond formed by the electrical attraction between positive and negative ions (cations and anions).

**Isotopes** Atoms having the same number of protons but different numbers of neutrons.

**Ketone** An homologous series containing a $C=O$ group within a carbon chain.

**Le Chatelier's principle** If a system at equilibrium is subjected to a change, the equilibrium tends to shift so as to minimise the effect of the change.

**Mass number (A)** The number of protons + the number of neutrons in the nucleus of an atom.

**Molar mass** The mass of one mole of a substance.

**Molar volume ($V_m$)** Volume occupied by one mole of a substance at a given temperature and pressure.

**Molecular formula** Shows the actual number of atoms of each element present in the molecule. It is a simple multiple of the empirical formula.

**Molecular ion** The positive ion formed in a mass spectrometer from the whole molecule.

**Monomer** A small molecule that can be made into a polymer.

**Non-renewable resources** Natural resources that cannot be reformed in a reasonable timescale.

**Nucleophile** A species with a lone pair of electrons that can be donated to an electron-deficient species.

**One mole** The amount of any substance that contains the same number of particles as there are atoms in exactly 12 g of carbon-12.

**Oxidation number** The number of electrons that need to be added to (or taken away from) an element to make it neutral.

**Ozone layer** A layer surrounding the earth that contains $O_3$ molecules.

**Percentage (%) yield** Percent yield = mass (or moles) of product obtained × 100% ÷ maximum theoretic mass (or moles).

**Polar bond** Has one end of the bond with a slightly positive charge and the other end with a slightly negative charge i.e. a dipole. The molecule is neutral overall.

**Polymerisation** The joining of a very large number of monomer molecules to make a long chain polymer molecule.

**Position of equilibrium** The proportion of products to reactants in an equilibrium mixture.

**Principle of conservation of energy** Energy cannot be created or destroyed, only changed from one form to another.

**Propagation** The reaction by which the process continues/grows.

**Radical** A species with an unpaired electron.

**Rate of reaction** The change in concentration of a reactant or product per unit time.

**Reaction mechanism** Shows the stages by which a reaction proceeds.

**Reflux** A process of continuous evaporation and condensation.

**Relative atomic mass** The average mass of one atom of the element relative to one-twelfth the mass of one atom of carbon-12.

**Relative formula mass** The sum of the relative atomic masses of all atoms present in its formula.

**Relative isotopic mass** The mass of an atom of an isotope relative to one-twelfth the mass of an atom of carbon-12.

**Repeat unit** The section of the polymer that is repeated to make the whole structure.

**Reversible reaction** A reaction that can go in either direction depending on the conditions.

**Salt** Compound formed when a metal ion replaces the hydrogen ion in an acid.

**Saturated compound** A compound in which all the C to C bonds are single bonds.

**Shielding effect or screening** The repulsion between electrons in different shells. Inner shell electrons repel outer shell electrons.

**Shortened formula** Shows the groups in sufficient detail that the structure is unambiguous.

**Skeletal formula** Shows the carbon/ hydrogen backbone of the molecule as a series of bonds with any functional groups attached.

**Standard solution** A solution for which the concentration is accurately known.

**Stoichiometry** The molar relationship between the amounts of reactants and products in a chemical reaction.

**Strong acid** Acid that fully dissociates in aqueous solution.

**Structural isomers** Compounds with the same molecular formula but with different structural formulae.

**Substitution reaction** A reaction in which one atom/group is replaced by another atom/group.

**Successive ionisation energies** A measure of the energy needed to remove each electron in turn until all the electrons are removed from an atom.

**Termination** The reaction that ends the process.

**Unsaturated compound** Contains C to C multiple bonds.

**Van der Waals forces** Dipole–dipole or temporary dipole–temporary dipole interactions between atoms and molecules.

**Volatility** How readily a substance vaporises.

**Wavenumber** A measure of energy absorbed used in IR spectra.

**Weak acid** Acid that partially dissociates in aqueous solution.

**α-particles** Nucleus of 2 protons and 2 neutrons, with positive charge.

**β-particles** Fast moving electrons, with negative charge.

**γ-rays** High energy electromagnetic radiation, with no charge.

**π bond** A bond formed by the sideways overlap of p orbitals.

**σ bond** Is made by end to end overlap of s or p-orbitals..

# Index

# THE PERIODIC TABLE

## Group

**Key**

```
    A_r      ← relative atomic mass
  Symbol
   Name
    Z       ← atomic number
```

s block — Groups 1, 2
d block
p block — Groups 3, 4, 5, 6, 7, 0
f block

| Period | Group 1 | Group 2 | d block | | | | | | | | | | | | Group 3 | Group 4 | Group 5 | Group 6 | Group 7 | Group 0 |
|---|---|---|---|---|---|---|---|---|---|---|---|---|---|---|---|---|---|---|---|---|
| 1 | 1.01 H Hydrogen 1 | | | | | | | | | | | | | | | | | | | 4.00 He Helium 2 |
| 2 | 6.94 Li Lithium 3 | 9.01 Be Beryllium 4 | | | | | | | | | | | | | 10.8 B Boron 5 | 12.0 C Carbon 6 | 14.0 N Nitrogen 7 | 16.0 O Oxygen 8 | 19.0 F Fluorine 9 | 20.2 Ne Neon 10 |
| 3 | 23.0 Na Sodium 11 | 24.3 Mg Magnesium 12 | | | | | | | | | | | | | 27.0 Al Aluminium 13 | 28.1 Si Silicon 14 | 31.0 P Phosphorus 15 | 32.1 S Sulfur 16 | 35.5 Cl Chlorine 17 | 40.0 Ar Argon 18 |
| 4 | 39.1 K Potassium 19 | 40.1 Ca Calcium 20 | 45.0 Sc Scandium 21 | 47.9 Ti Titanium 22 | 50.9 V Vanadium 23 | 52.0 Cr Chromium 24 | 54.9 Mn Manganese 25 | 55.8 Fe Iron 26 | 58.9 Co Cobalt 27 | 58.7 Ni Nickel 28 | 63.5 Cu Copper 29 | 65.4 Zn Zinc 30 | | | 69.7 Ga Gallium 31 | 72.6 Ge Germanium 32 | 74.9 As Arsenic 33 | 79.0 Se Selenium 34 | 79.9 Br Bromine 35 | 83.8 Kr Krypton 36 |
| 5 | 85.5 Rb Rubidium 37 | 87.6 Sr Strontium 38 | 88.9 Y Yttrium 39 | 91.2 Zr Zirconium 40 | 92.9 Nb Niobium 41 | 95.9 Mo Molybdenum 42 | 98.9 Tc Technetium 43 | 101 Ru Ruthenium 44 | 103 Rh Rhodium 45 | 106 Pd Palladium 46 | 108 Ag Silver 47 | 112 Cd Cadmium 48 | | | 115 In Indium 49 | 119 Sn Tin 50 | 122 Sb Antimony 51 | 128 Te Tellurium 52 | 127 I Iodine 53 | 131 Xe Xenon 54 |
| 6 | 133 Cs Caesium 55 | 137 Ba Barium 56 | 139 La Lanthanum 57 ▲ | 179 Hf Hafnium 72 | 181 Ta Tantalum 73 | 184 W Tungsten 74 | 186 Re Rhenium 75 | 190 Os Osmium 76 | 192 Ir Iridium 77 | 195 Pt Platinum 78 | 197 Au Gold 79 | 201 Hg Mercury 80 | | | 204 Tl Thallium 81 | 207 Pb Lead 82 | 209 Bi Bismuth 83 | (210) Po Polonium 84 | (210) At Astatine 85 | (222) Rn Radon 86 |
| 7 | (223) Fr Francium 87 | (226) Ra Radium 88 | (227) Ac Actinium 89 ▲▲ | | | | | | | | | | | | | | | | | |

▲ Lanthanoid elements
▲▲ Actinoid elements

**Lanthanoid elements**

| 140 Ce Cerium 58 | 141 Pr Praseodymium 59 | 144 Nd Neodymium 60 | (147) Pm Promethium 61 | 150 Sm Samarium 62 | (153) Eu Europium 63 | 157 Gd Gadolinium 64 | 159 Tb Terbium 65 | 163 Dy Dysprosium 66 | 165 Ho Holmium 67 | 167 Er Erbium 68 | 169 Tm Thulium 69 | 173 Yb Ytterbium 70 | 175 Lu Lutetium 71 |
|---|---|---|---|---|---|---|---|---|---|---|---|---|---|

**Actinoid elements**

| 232 Th Thorium 90 | (231) Pa Protactinium 91 | 238 U Uranium 92 | (237) Np Neptunium 93 | (242) Pu Plutonium 94 | (243) Am Americium 95 | (247) Cm Curium 96 | (245) Bk Berkelium 97 | (251) Cf Californium 98 | (254) Es Einsteinium 99 | (253) Fm Fermium 100 | (256) Md Mendelevium 101 | (254) No Nobelium 102 | (257) Lr Lawrencium 103 |
|---|---|---|---|---|---|---|---|---|---|---|---|---|---|